Japan's Rice Problem

Japan's Rice Problem

Masayuki Ogawa

Translated from the original Japanese by Makito Noda

CABI

CABI is a trading name of CAB International

CABI
Nosworthy Way
Wallingford
Oxfordshire OX10 8DE
UK

Tel: +44 (0)1491 832111
E-mail: info@cabi.org
Website: www.cabi.org

CABI
200 Portland Street
Boston
MA 02114
USA

T: +1 (617)682-9015
E-mail: cabi-nao@cabi.org

Originally published in Japanese under the title *Nihon no Kome Mondai by* Masayuki Ogawa 2022 under the label Chuko Shinsho.

A catalogue record for this book is available from the British Library, London, UK.

ISBN-13: 9781800627079 (hardback)
9781800627086 (ePDF)
9781800627093 (ePub)

DOI: 10.1079/9781800627093.0000

Commissioning Editor: Ward Cooper
Editorial Assistant: Emma McCann
Production Editor: Rosie Hayden

Typeset by Exeter Premedia Services Pvt Ltd, Chennai, India
Printed and bound by CPI Group (UK) Ltd, Croydon, CR0 4YY

Contents

Preface for the English version of *Japan's Rice Problem*

This book is an English translation of my June 2022 book, *Nihon no Kome Mondai* (*Japan's Rice Problem*, Chūō Kōron Shinsha, Tokyo).

Alhough rice is grown all over the world, conditions surrounding rice farming differ greatly from country to country. This fact is one unique feature of rice production. As countries develop economically, they typically experience a dietary transition similar to the phenomenon seen in developing countries where traditional diets high in cereal and fiber shift to more Western-style diets high in animal-source foods. This trend has been particularly notable in Asian countries, leading to a serious decrease in the demand for rice in some countries. In Africa, in contrast, rice production has spiked greatly in the past ten years, a turn of events that can be attributed to the intense promotion of rice farming in order to accomplish self-sufficiency in cereals.

Globally speaking, Japan has actually experienced the following five states of affairs ahead of other countries: (1) self-sufficiency in rice, (2) being first in economic development among countries where rice cultivation flourishes, (3) reduced rice consumption, (4) various rice-related policies including production control, and (5) the handling of issues related to food, agriculture, rural areas, and the environment accompanying reductions in rice production. Rice has occupied a special position in Japan's economy, society, and culture throughout its history.

This book aims at transmitting Japanese academic insight into rice to the world, including the history of pertinent Japanese policies, to promote overseas understanding of Japan and to offer observations that may be useful to experts and policy authorities worldwide.

While there have been countless studies on rice in Japan, many of them are full of occurrences and terms that can be understood only by experts in Japan. While I am proud to say that in writing the original Japanese-language book, I refrained from using jargon as much as possible, I sincerely hope that readers of this English edition will find it an easy-to-read yet highly professional book on the history and status of rice in Japan. It is my hope that this book becomes a basic document on Japan's rice to be read all over the world for a long time to come.

A unique feature of this book is that it criticizes Japan's failure to strategically formulate a future outlook for rice and rice paddies even before the achievement of self-sufficiency in rice. I like to think that this criticism offers an important viewpoint, not only for Japan, which has already experienced declines in the demand for rice and in population growth, but also for those present-day developing countries that are currently struggling to increase rice production to secure food as well as countries where declines in the demand for rice are predicted as economic development advances.

This English-language edition was professionally translated by Makito Noda. While it is a faithful translation of the Japanese original, in some portions of the book sentences and paragraphs have been reordered or restructured to make the writing and messages of the original easier to understand for overseas readers. Needless to say, the entire translated manuscript was approved by myself and benefited from the professional expertise of the editors at CAB International, the publisher.

Readers should be informed that my proposal concerning amendment of the Basic Act on Food, Agriculture and Rural Areas – which is regarded as the constitution of Japan's agricultural policy – which I outline in the section of Chapter 7 entitled Enacting a new Basic Act and Constitution, has already been realized. Upon Prime Minster Kishida Fumio's directive for review of the act 3 months after publication of *Nihon no Kome Mondai*, the first amendment to the act was promulgated in June 2024, only 2 months prior to my writing these words. Therefore, this book does not contain discussion of the amendment.

I would like to take this opportunity to express my deepest gratitude to Mr Noda and the experts at CAB International who supported me in the publication of this book. I am also grateful for the immensely beneficial advice provided by Messrs Mark Gresham and Ivan Cardenas.

Masayuki Ogawa
August 2024

Introduction

Rice has long been a staple in Japanese families. Yet in 2010, the annual purchase of bread by household surpassed that of rice. For the next ten years, the average annual household spending on rice continued to decline, reaching only ¥18,503 by 2020. This amount is slightly above the average household spending for noodles (¥16,510) but far below that for bread (¥25,551). The average Japanese eats as little as 139 g of rice a day – which, when cooked, barely fills two rice bowls – even including meals eaten out and ready-made precooked meals purchased to be eaten at home and/or the office. In terms of an energy source, rice occupies merely 21% of the total Japanese caloric intake.

In the past, rice was the source of almost half of the Japanese daily caloric intake. Many readers of Miyazawa Kenji's 1931 poem *Ame ni mo Makezu* (*Be Not Defeated by the Rain*) may find his reference to 'dining daily on four cups of brown rice...' surprisingly voluminous. One hundred years ago in 1922, the average Japanese ate 3.3 times more rice per person than today. Even during the food shortage in the period immediately after World War II, Japanese ate twice as much rice as today. Although the Japanese continued to consume 2.3 times more rice than today in 1962 in the midst of high economic growth, after that the volume of rice intake began to decline. A dwindling population further reinforced the declining trend, resulting in an annual decrease of some 100,000 metric tons of domestic demand for rice for food in recent years. In 2021, the amount of land for rice paddies to produce rice for food in Japan decreased by 63,000 ha from the previous year.

In terms of area, the rice paddies into which earlier generations poured their wisdom, capital, and labor to grow more and better rice has shrunk to 70% of the peak period. Today, only 58% of surviving rice paddies are used to grow rice. The typical pastoral scenes in Japan of endless stretches of rice paddies filled with water in the spring, lush with green rice plants awash in the buzzing of cicadas and a chorus of frogs in the summer, and full of ripe golden ears in the autumn are quickly becoming things of the past.

On the other hand, in no other period of Japan's history has the competition among rice farmers and rice-producing regions been so brisk and fierce. The days of absolute reign by the Koshihikari brand are long gone. Today, many varieties of rice, including such brands as Tsuyahime and Yumepirika as well as rice for business use (in restaurants and for ready-made meals), such as Mitsuhikari, Akidawara, and Moeminori, have entered the market to fight it out in fierce competition. Yet even this situation is no longer a competition among the central players of the Japanese economy. Rather, the brands in this increasingly severe competition aim to gain a larger share in a shrinking market. The very survival of rice production in Japan is at stake, giving this competition an air of pathos. The rice farmers who have chosen to produce rice for export are faced with such hardships as high transportation costs, steep middle margins, and fierce competition from already plentiful overseas rice producers.

Moreover, production of rice for food is also facing competition from rice for rice flour and animal feed, both of which are increasingly in high demand. In fact, rice for processing is in such high demand that in 2016 the National Council of Organizations of Demanders of Rice for Processing, which wants expanded production of rice for confectionery goods and sake (Japanese rice wine), and four associations of animal feed manufacturers, which are in need of rice processed for livestock feed, separately made a highly irregular appeal to rice farmers to expand rice production for them.

Rice in Japan today is plagued by so many complicated problems, ranging from consumers' loss of interest in eating rice to an increasing number of abandoned rice paddies and ferocious competition in the dwindling rice market, that it is difficult to grasp the entire picture. This book anatomizes the rice problem in Japan through my own original perspective so as to unmask the essence of the issue and, in so doing, offer an explanation on how relations between the Japanese people and rice have drastically changed, with special attention on several turning points.

I'm aware that for some readers, particularly urban dwellers, 'Japan's rice problem' may not ring a bell. In fact, I myself hardly ever hear of someone suffering from Japan's rice problem, except in conversations with those engaged in agriculture in one way or another. While the increase of abandoned rice paddies accelerated by loss of consumer interest in rice and the ageing of rice farming families are problematic, they do not themselves cause such societal problems as rice supply shortages or hikes in rice prices. On the contrary, fierce competition among rice brands has allowed consumers to enjoy sampling various rice brands from different regions. Today's consumers are spared the trouble of cooking rice when they purchase ready-made rice balls made from any of the many brands of rice or pre-packaged cooked rice. Almost all the rice sold in Japan is domestic, assuring consumers a safe and stable supply. This sense of security is further reinforced by the Japanese government's stockpile of rice. Business efficiency and limited marketing channels are indeed problems for Japan's rice industry, but these fall within the boundary of ordinary corporate efforts. In fact, there are plenty of industries which are suffering much more seriously from these problems than the rice industry. One can even say that, from the viewpoint of the Japanese consumer, there is no commodity that is more trouble-free than rice.

Just because Japanese consumers are not suffering when it comes to rice by no means indicates that there is no rice problem in Japan. It is precisely because all the rice-related problems have been concentrated in a single, yet-to-be-attended-to problem that people in Japan can go on enjoying carefree daily life. This book presents a deep exploration into the true nature of the rice problem. It is my strong wish that those people for whom the rice problem is not familiar and those people who feel they are not affected at all by the rice problem read this book.

I believe that when the Japanese people find out the truth about the unattended, unsolved problem revealed in this book, they will realize that the rice problem troubling Japan today is only a prelude to an imminent, unprecedented crisis.

Note

The major reason any discussion about rice is hard to comprehend is the abundance of terminology and government policies/laws. The same terminology may mean different things in different times, thus further complicating things and making it hard even to memorize relevant terms. This is why I decided to refrain from using terminology and policy/law names as much as possible.

In order to prioritize understandability over definitional precision, I also decided to ignore or simplify minor differences in meaning. For instance, in the Japanese language, 食料 and 食糧 share the same pronunciation: *shokuryō*. But the former means food as a whole, while the latter represents basic foodstuffs, most typically staple food such as rice, wheat, and corn. Strictly speaking, these two words should be differentiated, but in this book, I decided to integrate the two.

Instead of the phrase *net food supply*, which includes wasted food, this book uses the wording *the amount eaten*. The rice farming calendar includes the terms *product year* and *rice year* in addition to the ordinary *(calendar) year* and *fiscal year*. This book renders all such references as calendar years as long as no substantive difference is evident.

Not all rice paddies are planted with rice. This is a major characteristic of Japanese agriculture. Therefore, this book uses the term *active rice paddy* to refer to paddies where rice is planted. Other paddies may be barren of rice or planted with other, non-rice crops.

This book deals with three types of rice: (1) rice to be cooked and eaten either at home or other eateries; (2) rice to be processed into other rice products such as rice wine (sake), rice paste (miso), and rice crackers (*senbei*); and (3) rice feed to be fed to animals. Unless otherwise specified, the word *rice* in this book means the first kind of rice to be cooked and eaten either at home or other eateries, including rice purchased as precooked meals.

1 Separating the Rice Issue and the Rice Paddy Issue for Better Understanding

Abstract

Chapter 1 analyzes and expounds the current situation of rice production in Japan, including the status of the rice cultivation business as well as the reasons behind the small amount of rice imports. Pointing out that Japan's rice problem should be divided into issues centered around rice per se, on the one hand, and those associated with rice paddies, on the other, this chapter posits that in Japan, where the supply of rice is sufficient, the rice paddy problem has become a salient issue. Japan's rice production is essentially for domestic consumption and is unique in the sense that, if rice is produced in all the rice paddies in Japan, supply will grossly supersede domestic demand. The chapter also expounds the impending risks should Japan cease to produce rice, ranging from the decay of scenic beauty in rural areas to liability to flood damage.

Familiar Rice vs Little-known Rice Production

Fact: 3.6 per cent of the national land is used to produce rice

Being so frequently served, rice is a highly familiar food for Japanese people. Even those who have never experienced rice planting or rice reaping will have seen a rice paddy or rice plants growing at least once in the past year. The Japanese archipelago is surrounded by the sea, and two-thirds of the land is forested. Thus, as soon as one leaves densely populated cities full of buildings, the sea and mountains come into sight whether one likes it or not. The same is true for rice paddies: they come into view without any conscious effort. Does this mean that rice paddies are highly visible because they are as ubiquitous as the sea and mountains? No, it doesn't.

In 2020, rice paddies occupied 2.4 million ha in Japan, of which 1.37 million ha– about the size of Fukushima prefecture– were devoted to the cultivation of rice for meals. This is merely 3.6% of the entire national land mass, an amount that can be compared to a stamp on a postcard. It is because people and rice are packed together in narrow pieces of inhabitable land that people feel that rice paddies are everywhere in Japan.

Once, I asked my students in the faculties of agriculture and economics to tell me what rice paddy land was as a percentage of the total national land area within 5 s. Interestingly, students from urban areas tended to be more accurate, citing 1–5%, while many of those from rural regions guessed 10–15%, much higher than the reality. When I asked the same question of rice farmers during the course of a survey on farm households, many answered 15% or 20%. Asked why they came up with these percentages, many farmers said that rice paddies are everywhere in their hometowns and that Japan is full of rice paddies. Even though, admittedly, only a small number of people answered the question, it was still surprising that rural dwellers, to whom rice paddies are a very familiar sight, and rice farmers, who are experts in rice farming, made such wrong guesses. Pondering on why, I realized that this tendency can be explained when one looks closely at the sphere of life.

Here I define sphere of life as the national land excluding forests, wilderness, rivers, and other difficult-to-access areas. Based on this definition, let us look into the land use situation in Japan (see Fig. 1.1). In Tokyo, Kanagawa, and Osaka prefectures, where 31.6 million people, or a quarter of the population, reside, paddy fields occupy only 3.1% of the space. The rate goes down to 2.3% when limited to active rice paddies alone (where rice is actually planted). Strangely

DOI: 10.1079/9781800627093.0001

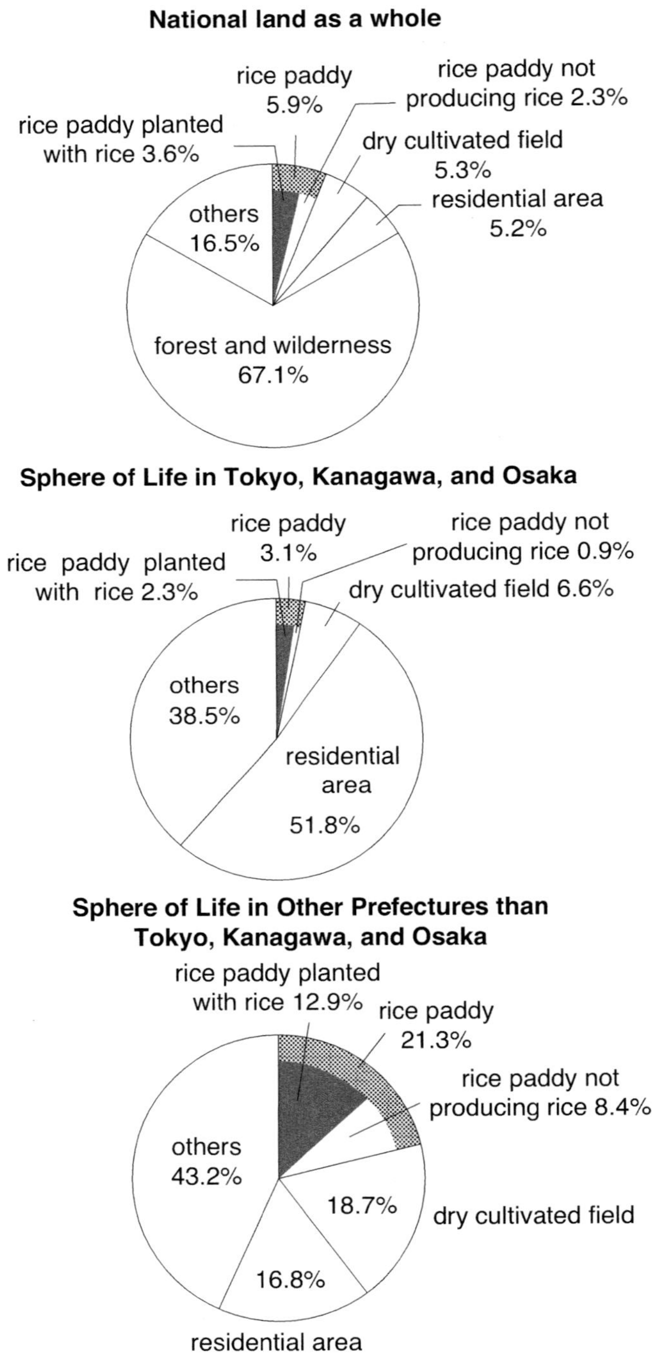

Fig. 1.1. Status of national land use, 2016. *Source*: based on the Ministry of Land, Infrastructure, Transport and Tourism's *Tochi Shoyū/Riyō Jōkyō Chōsa* [*Survey of Status of Land Ownership and Use*] and the Ministry of Agriculture, Forestry and Fisheries' *Sakumotsu Tōkei* [*Statistical Survey on Crops*].

enough, these rates are quite proximate to those for paddy fields and active rice paddies in Japan as a whole: 5.9% and 3.6%, respectively.

On the other hand, in the remaining 44 prefectures, where 95.33 million people, or three-quarters of the population, dwell, paddy fields occupy 21.3% of the sphere of life, with the rate for active rice paddies being 12.9%. Rates for paddy fields and active rice paddies rise when Hokkaidō, where there are more cultivated fields than other prefectures, is excluded. Thus, if rice farming and paddy fields in one's own sphere of life immediately come to mind when asked about the percentage of rice paddies, it is not surprising at all that people wrongly guess the percentage of rice paddy in the total national land area to be 15% or even 20%.

Rice problem: urban areas vs rural areas

Often, discussions on the rice problem in Japan go nowhere because people are arguing on different planes. The major reason behind this, as we've just seen, is the perception gap even among the Japanese themselves concerning the presence of rice farming. Thinking in the extreme, should rice farming become non-existent in Japan, the impact on day-to-day scenery and the sphere of life will differ between urban and rural areas, even though the unavailability of domestic rice will affect them equally. While urban people may not be seriously affected, no rice farming would mean a complete change in the use of a vast expanse of land – almost equivalent to all residential areas nationwide – for the rest of Japan. The impact would be grave.

The meaning of the rice problem differs significantly whether one focuses only on rice as a food or includes the issue of rice paddies. One's psychological proximity with rice farming also plays a role. Today, rice farming is no longer a usual scene in Japan. Due to the falling birthrate and the concentration of population in urban areas, one out of ten children reside in Tokyo today. At the same time, more than half of all municipalities nationwide have become under-populated. Thus, the number of people who can visualize Japan's authentic scenery, including rice farming, is also decreasing. In this way, rice farming is becoming increasingly psychologically remote to the Japanese people.

Paddies growing rice come into view once a traveler is out of town. This very common sight in Japan is quickly becoming a thing of the past. (Photographer: Joe Maruoka)

Rice farming is wondrous

One wonders how rice has come to have such a close existence with the Japanese people. The primary among several possible explanations is that rice is a highly fragile alien species. The rice plant was introduced to Japan from abroad as a food source. Unlike such other alien species as black bass, racoons, or tall goldenrod, however, rice is not endowed with a tough vital force: even 3000 years after its introduction to Japan, rice has not yet indigenized. It may be trite to quote the popular saying that the Chinese character for rice symbolizes the 88 forms of care that rice requires to grow, but there is truth here. Unattended, the rice plant does not sprout and grow vigorously on its own. Once farmers stopped tending rice paddies, rice would become extinct in no time. Japan's rice exists only within the Japanese sphere of life. And it is to this rice that the Japanese people have devoted their energy and efforts. For instance, take the tiny paddy fields found between remote mountains, the sight of which never fails to amaze visitors. The neatly piled stone walls supporting layers of rice terraces betray the extraordinary enthusiasm farmers of long ago poured into rice farming. In flat areas, the sight of rice paddies meticulously lined up like a chessboard covering a vast plain is equally spectacular.

The fact that rice has been harvested everywhere in Japan tells us that it must be an extremely popular crop. Its attractiveness as a crop with an exceptionally high yield is the second reason why rice has come to have such a close existence with the Japanese people. Even before great breed improvements, rice is much more capable of feeding a population than, for instance, wheat; so much so, in fact, that Adam Smith, who is known as the father of modern economics, made special reference to it in his *Wealth of Nations*.

When one plants a grain of rice, how much rice does it yield? This is the question numerous farmers as well as researchers have tried to answer, using all kinds of species, methods, and calculations. Today, it is believed that, generally, a grain of rice yields 500–1000 times more grains. This means that when about ten grains of rice grow healthily, about 6500 grains, or about one *go* (150 g) of rice, can be obtained. Historically, the multiplying factor of rice was 30 times in the Edo period (1603–1868) and 10 times even in the Nara period (710–794). Compared to that of wheat, which had a multiplying factor of three times in medieval Europe and 10–20 times even today, it has to be said that the productivity of rice is amazingly high.

The third reason rice has come to have such a close existence with the Japanese is its excellent nutritional value. Carbohydrate, rice's main ingredient, is easily absorbed by the body. Rice provides protein, minerals, and vitamin C. Brown rice, in particular, contains plenty of vitamin B1 (or thiamin), fiber, niacin, folic acid, and lipid. Almost all kinds of amino acids, which are the building blocks of protein, are found in rice. Until the 1970s, Japanese relied on rice for one-quarter to one-third of their protein intake, which allowed them to be starvation free even without fish or meat. Rice also contains all of the essential amino acids in a better balance than wheat (better balance means a higher amino acid score), and when cooked, rice produces oligosaccharide, which improves the intestinal environment.

On top of these merits, Japan's pluvial climate and volcanic acidic soil are suitable for rice, which is relatively resilient to acid and takes a lot of water to grow. Rice requires a lot of care as well, but it can feed a large population. As rice growing became widespread in Japan in the Jōmon period (between *c.*14,000 and 1,000 BCE), so too did Japan's population begin to grow, starting from the western region of the country.

Rice farming also promoted domiciliation, contributing to organization of human settlements and towns. Furthermore, rice is useful as a raw material for sake (rice wine) and miso (bean paste), while rice straw can be processed into such items for daily life as straw slippers and straw rain capes. Rice straw can also be mixed with building materials for walls. In short, the Japanese have been supported by rice for food, clothing, and housing.

Let's now turn our eye to the rest of the world. A high-yielding rice breed called IR8 appeared in 1966, triggering the so-called Green Revolution to help feed large numbers of people in Asian developing countries. In Africa, a new breed of hybrid rice, NERICA (New Rice for Africa), was developed in 1994. Today, rice farming has spread to every corner of the world. Of the world's three major grains (rice, wheat, and corn), rice is the most widely eaten. Rice is the staple food for two-thirds of the world's entire population.

Overall, 500 million tons of rice is produced annually in the world, of which 52% is produced by the top world rice producers: China and India (note that use of 'tons' means metric tonnes throughout the book). The list of top 20 producers includes Asian countries – Japan is no. 12 – as well as countries in Africa and the Americas. Today, Brazil produces more rice than Japan, and so did the United States in 2010 and 2020 (see Fig. 1.2).

Rice farming supported by water, forests, and mountains

Different agricultural crops require different nutrients. When the same crop is repeatedly grown in the same plot, the soil's nutritional balance is lost. Moreover, this practice attracts certain species of pests and microorganisms, causing a field's yield to drop. This phenomenon is called 'injury by continuous cropping.' While rice also causes injury by continuous cropping when grown in a dry field, damage hardly occurs when rice is grown in (wet) rice paddies where the soil is washed by water. This is the greatest

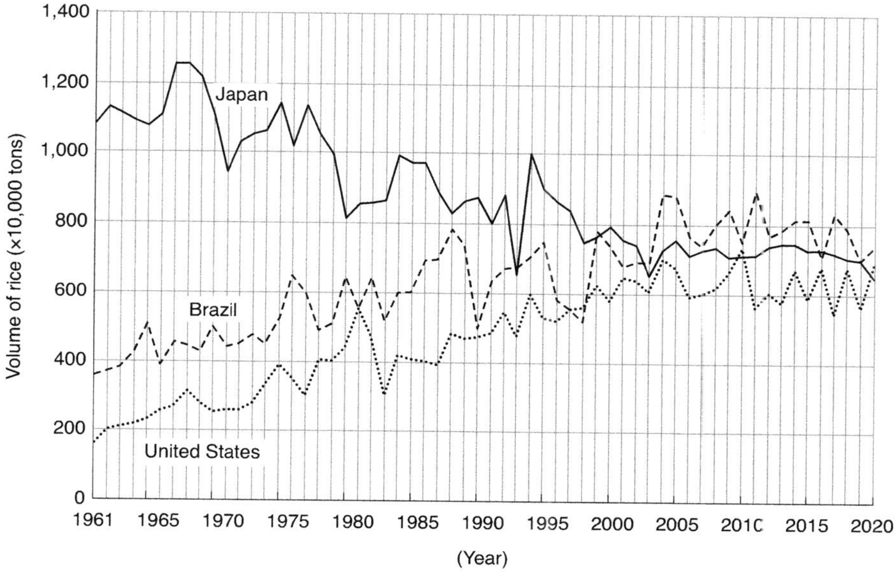

Fig. 1.2. Changes in volumes of rice produced in Japan, the United States, and Brazil. *Source*: based on the Food and Agriculture Organization Statistics (FAOSTAT) database of the United Nations.

reason that rice has been grown without a break in Japan, where almost all rice is grown in water-filled rice paddies.

Water brings with it rich nutrients collected during its passage through the mountains to the rice paddies. Rice paddies filled with water produce allow cyanobacteria to grow, making the paddy a cordial environment for absorbing nitrogen in the air and converting it to fertilizer. Another merit of wet rice paddies is that the amount of weeds is much lower than for dry fields. By intentionally draining the paddies of water occasionally, rice farmers can damage the seed of hydrophilic weeds. In addition, water can keep rice paddies warm in the cold early days of spring and lower paddy temperature with free-flow irrigation in the summertime.

Mountains in Japan are not only sources of water but also wintertime reservoirs for snow. Mountains protect against rice blast, an arch enemy of rice, by sending mild winds to rice paddies. It has been long recognized that mountains affect the growth of rice. Folk legend has it here and there in Japan that in springtime, the mountain god morphs into the rice paddy god to produce a rich harvest, whereas, in autumn, the rice paddy god transforms itself back into the mountain god. The importance of maintaining mountains' function as a water resource by preserving forests has been known for a long time. Rice farming in Japan is a major industry nurtured by mountains, forests, and rivers, which combined occupy 70% of the national land. Of the total 80 billion m^3 of water used annually in Japan, including that for daily home life and industrial use, two-thirds is used in rice paddies.

Rice paddies as dams

In times gone by when there was imminent danger of flooding, communities in Japan relied on the emergency method of deliberately opening barrage gates so as to guide river water to the rice paddies. This saved residents and their homes from flood damage. Cases of rice paddies hit by floods are abundant everywhere in Japan. Particularly well known from the early Edo period are the large-scale public works to divert floods to rice paddies by Naridomi Shigeyasu of the Saga-*han* and Katō Kiyomasa, lord of the Kumamoto-*han*. On the societal consensus that people's day-to-day life relied on the sacrifice of rice paddies during floods, some regions in Japan used to reduce the annual land tax for farmers who were flood victims, and unaffected farmers pitched in to work to restore damaged paddies.

In contrast to upland field crops such as soy and other vegetables, which are vulnerable to flood damage, the rice plant hardly incurs damage even when it is fully immersed in water overnight. Unless the immersion coincides with the day-long blooming/pollination period or with the fruition period of ears of rice, the harvest is not seriously affected. For Japan, which is deluged with torrential rains and floods, rice farming has been a reliable method for securing a stable food supply, despite the climate.

Although rice paddy water is only 10 cm deep, the vast areas occupied by rice paddies mean that paddies can store significant amounts of water. It is calculated that some 280,000 ha of rice paddies along the Tonegawa river can contain a maximum 280 million m^3 of water, which is about 60% of the total storage capacity of eight dams on the Tonegawa riverine system (480 million m^3). Even if the river overflows, the damage will not be too serious as long as the overflow is directed to the rice paddies.

Today, however, this traditional failsafe system for handling the danger of a flood beyond the capacity of dams is no longer as reliable as before. This is because in many places rice paddies have been converted to residential lots. Now that rice farming is no longer close to the Japanese heart, risks from flooding disasters have heightened. To combat this, there has been a gradual rise throughout the country in the number of 'rice paddy dams,' or rice paddies equipped with drainage outlets to enhance their function as barrages.

Much of the Rice Problem is Actually a Rice Paddy Problem

Real nature of the rice problem

Each period of Japan's history had its own rice problems. They included, for instance, food securement, prices, gluts, import bans, tariffs on imported rice, and deregulation. Common among all of these problems in the past is that the focus of discussions was rice per se. In contrast, today's rice problem does not share this feature.

At the time of publication of this book, Japan is more than capable of producing enough rice and, therefore, a dearth of rice is not feared. Rice farming is no longer subsidized in Japan. Except for specially designated insurance for rice farmers and rice, there is basically no government subsidy for farming rice for meals. And there is no government policy that prohibits rice farming. Farmers' efforts toward tending paddies and managing their farming business are directly reflected in their business results. An increasing number of improved rice species have been developed, and rice farming to meet new needs has also intensified. As new rice species are developed and as global warming progresses, it has become possible to produce rice even in Hokkaidō, which used to be a barren land for rice. Today, brand rice farming is quite popular in this northern island. As of 2021, nationwide there were as many as 893 brands of rice for meals, 137 brands of glutinous rice, and 224 brands of rice for brewing sake. There are also numerous combinations of rice brands and producing regions. Rice farming today is, so to speak, in an age of rival brands fighting for their very survival. There once was a period when rice was enshrined as a holy sanctuary to be protected from the rough waves of international trade. That time has passed now that the protective barrier has been penetrated by such international arrangements as the Trans-Pacific Partnership (TPP), allowing Japan to both import and export rice. In the competition of today, participants in the rice farming industry are very conscious of the global standard.

Today's rice problem in Japan would not be much of an issue if rice as a material or a commodity was the only point to consider. What complicates the rice problem is the financial burden on the Japanese government to indirectly support rice farmers. Most representative of this burden is the massive amount of financial support for production of non-rice agricultural products in rice paddies not used to grow rice, as well as for the maintenance and improvement of active rice paddies. While it is often said that rice imposes a heavy financial burden on the government, it is, strictly speaking, support for rice paddies that is the heavy burden. Saying this now is to get ahead of ourselves a little regarding the core argument of this book, but in my view today's rice problem in Japan is no longer a problem of rice per se. Rather, the rice problem has already been displaced by a rice paddy problem. Because prompt resolution of the rice problem was given priority, the rice paddy problem has been left unsolved, relatively speaking, and has now grown into something too large to be ignored any longer.

Today, much of what is referred to as rice problems are actually problems related to rice paddies. By separating the issue of rice from the issue of rice paddies instead of mixing the two up, one can see the essence of the current rice problem in Japan.

In the old days when most rice paddies were used to farm rice, issues with rice paddies were automatically discussed when the rice problem was discussed. These days, however, even when aspects of the rice problem, such as raising the profitability of rice farmers or finding successors to rice farmers, are improved, that does not automatically solve issues related to rice paddies, such as what to do with rice paddies where rice is not produced. Once it was sufficient to focus on rice to discuss the rice problem. That time is long gone, and we can no longer discuss the true nature of the rice problem unless attention is paid to the issue of rice paddies, because that is the root of Japan's current rice problem.

Rice paddy excess deserves more attention

When talking about the rice problem in Japan, such terms as *rice glut* and *rice price drop* may come to mind. In recent years, for instance, domestic demand for rice declined due to the COVID-19 pandemic. During the pandemic, many people chose to cook their own meals at home because they were working from home. This should have increased the demand for rice, but because the damage to the food service industry was so heavy, the amount of purchased rice actually declined drastically (including rice categorized as food loss). Accordingly, rice production was reduced by 224,000 tons in 2021. But the reduction proved to be insufficient. Tons of old rice remained stockpiled in 107 agricultural cooperatives' warehouses even when newly harvested rice of the year began to come onto the market.

The so-called rice glut is indeed a serious problem today. But, in my view, rice paddy excess – that is, there being more rice paddies than necessary to produce the amount of rice consumed – deserves more attention. Behind the rice glut is always this surplus rice paddy issue. The rice glut is only the tip of the iceberg that is the problem of surplus rice paddies (see Fig. 1.3).

The financial situation for farmers and farmland size

Hearing of Japan's rice problem, such characterizations as 'compact and money-losing rice farming' and 'heavily subsidized rice farming' may come to mind. Or perhaps one thinks of the news media reports of large-scale, high-income rice farmers as a business success. It seems that rice farmers in Japan are bound to be perceived in a stereotypical dichotomy: traditional rice farmers with poor business sense vs advanced rice farmers with excellent managerial sense. But is this portrayal really accurate? The difference in financial situations among rice farmers can be more clearly understood when, again, one focuses attention on rice paddies.

Figure. 1.4 shows the income of mainly rice paddy-tending farmers and the proportion of rice in their harvests arranged in order of farmland size. The larger the farmland, the more full time the farmers become and the smaller their income from side jobs or pensions becomes. Interestingly, the larger the farmland, the higher the proportion of agricultural subsidy in a farmer's total income. While Fig. 1.4 covers only a few of numerous agricultural subsidies, it still shows that the proportion of agricultural subsidies in total income is, on average, as high as 77% for farmers with more than 20 ha of farmland. This is far greater than the 0.6% for farmers with less than 0.5 ha. Also, the larger the farmland, the lower the proportion of rice in the total harvest. It can be summarized from this data, then, that larger farmers are inclined to rely more on agricultural subsidies and grow more crops other than rice. In contrast, smaller farmers tend not to rely on subsidies and to concentrate on rice farming. A more or less

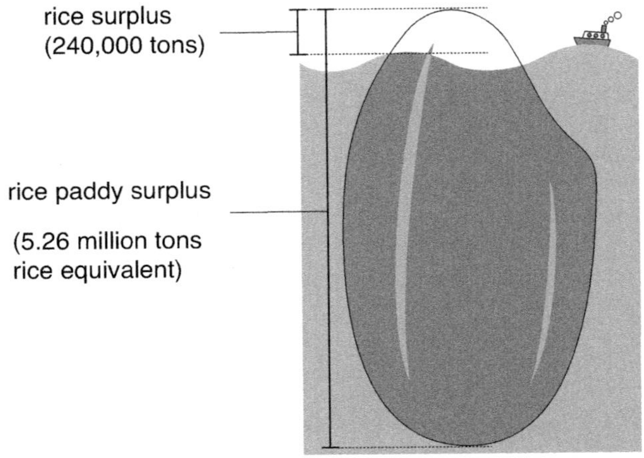

Fig. 1.3. The rice surplus is just the tip of the gigantic iceberg of the rice paddy surplus. *Source*: based on the Ministry of Agriculture, Forestry and Fisheries' *Sakumotsu Tōkei Chōsa* [*Statistical Survey on Crops*] and *Minkan Zaiko no Suii* [*Changes in Non-governmental Stockpiles*]. *Note*: for the sake of simplicity, it is assumed that all rice paddies yield the same amount of rice. The rice surplus is calculated as the change in the amount of non-governmental stockpiles at the end of September 2021 from the previous year.

similar tendency is also observed among incorporated agricultural firms (see Fig. 1.5).

Reasons behind different rice farming patterns

Figure. 1.4 shows that income from agriculture for small-scale farmers with less than 0.5 ha of farmland is below zero. That translates to a deficit of ¥391 per 5 kg of rice. Why, then, do those farmers continue to grow rice?

For one thing, the deficit incurred by rice farming can be absorbed, first, as a food expense when the harvested rice is consumed at home and, second, as a social expense when the rice is given to friends and relatives as a gift. According to the most recent Retail Price Survey compiled by the Statistics Bureau of Japan, Ministry of Internal Affairs and Communications, 5 kg of rice is sold at ¥2300. In other words, from the viewpoint of living expenses, it is still saves money to produce and consume the rice rather than buying it if the deficit remains below ¥2300 per 5 kg. A deficit of ¥391 is within a permissible range for a rice farmer with less than 0.5 ha of farmland. Besides, home-grown rice brings with it a sense of safety and satisfaction when it is eaten at home and a sense of premium value when it is given to friends and relatives. In addition, a family-run rice farm can be much more resilient in severe business environments by underestimating family members' wages at zero or, at least, the minimum wage level and interpreting their labor hours as exercise to stay physically and mentally healthy or as time pursuing a passion in life.

The significant difference seen in management styles among rice farmers is attributable not so much to the presence or absence of managerial sense, but chiefly to how they perceive rice paddies. That is, it is whether rice paddies are regarded as a venue to produce a commodity or a place to bring meaning into one's life by growing rice for consumption by family and friends.

Rice farming to preserve rice paddies

There are other reasons rice farmers continue to produce rice in spite of poor financial return.

Total area of rice paddy planted with agricultural produce	below 0.5 ha	5–7 ha	10–15 ha	over 20 ha
Proportion of agricultural income in total income	— in deficit	56% low →	73%	85% high
Proportion of agricultural subsidy in total income	0.6% low →	26%	46%	77% high
Proportion of area planted with rice in total rice paddy area planted with agricultural produce	99% high ←	89%	71%	56% low
Agricultural subsidy	¥25,000	¥1,382,000	¥3,497,000	¥11,725,000
Income from agriculture	–¥124,000	¥3,007,000	¥5,533,000	¥13,090,000
Income from non-agricultural sources	¥1,833,000	¥1,002,000	¥995,000	¥1,387,000
Pension, etc.	¥2,396,000	¥1,363,000	¥1,102,000	¥847,000
Total income	¥4,105,000	¥5,372,000	¥7,630,000	¥15,324,000

Fig. 1.4. Income of farmers heavily reliant on rice paddies and rice production. *Source*: based on the Ministry of Agriculture, Forestry and Fisheries' *2018-nen Einō Ruikeibetsu Keiei Tōkei (kobetsu keiei/ suidensaku)* [*2018 Statistics on Management by Farming Type (Individual Management/Rice Paddy)*] "Income from agriculture" is the amount obtained by subtracting expenses from income, including agricultural subsidies. "Total income" is the sum of the three amounts: "Income from agriculture," "Income from non-agricultural sources," and "Pension, etc." .

In today's rice farming, which relies heavily on machines, machinery-related expenses occupy as high as 20% of total expenses. More often than not, farmers use their machines long after the depreciation period ends as long as they are operative. Some farmers do not save enough money to buy new machinery. Thus, it is only after their machinery breaks down that they start looking for subsidies and exploring solutions such as commissioning other rice producers to harvest their paddies or sharing machinery with other farmers.

Many rice farmers choose to continue producing rice as long as they can even though their future business prospects are not bright. Motivations vary for holding on to rice farming out. As long as one possesses a rice paddy, one could benefit financially because there is always a possibility for the land to be claimed by the government to construct roads and railways or to be bought up by a private industrial company.

A sense of duty to protect the land inherited from their ancestors is an additional, deep-rooted motive for farmers to persist despite incurring a deficit. In the 1980s, the agricultural economist Soda Osamu found that farmers who gave up on farming would not part with all of their farmland. These farmers retained some 30 ares (3000 m^2; 1 are = 100 m^2) of land, which could produce subsistence-level crops for two families, given the technological level of the time. Soda called this phenomenon a 'theory of retention of 30 ares farmland.' Holding on to their rice paddies until the very end was, in a way an insurance for farmers against a bad economy, war, or extraordinary natural phenomena so as to secure their family's food as well as employment.

Concentrating only on rice itself, one can objectively calculate profits and losses for rice farming based on sales profit and the costs of production. Factoring in rice paddies, though, requires that the calculation also includes such subjective and trans-generational factors as

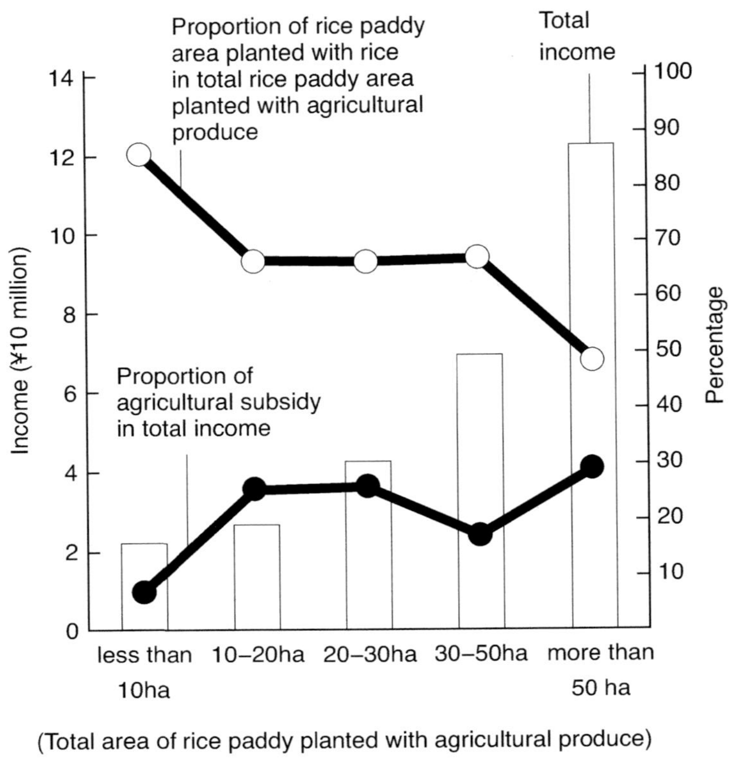

Fig. 1.5. Income of incorporated agricultural firms heavily reliant on rice paddies and the proportion of rice production in their income. *Source*: based on the Ministry of Agriculture, Forestry and Fisheries' *2018-nen Einō Ruikeibetsu Keiei Tōkei (kobetsu keiei/suidensaku)* [*2018 Statistics on Management by Farming Type (Individual Management/Rice Paddy)*].

assessment of the land as an asset, a source of pride as something inherited from ancestors, and an insurance for family and descendants.

Rice paddies surviving urbanization

Farmers continuing to produce rice despite incurring a deficit is seen even where urbanization has been pursued (the so-called urbanization promotion area, or UPA). Take the example of Mr A., who lives in a UPA in Gifu city. Mr A. harvested 471 kg of rice from his 10 ares (10,000 m^2) of paddies and sold them for ¥78,500. Subtracting necessary expenses and not counting his own labor, Mr A. still ran a deficit of ¥36,746. Because he had to pay a property tax of ¥215,000, his total deficit ballooned to ¥251,746. This deficit was offset by his real estate income from a property he owns in the commercial district and other side businesses.

Mr A. could easily improve his balance of payments from farming if he chose to grow Japanese radish or green soybean instead of rice, even though both may require heavy labor and/or extra care. Still, Mr A. continues to produce rice. Why? Because, aside from the convenience of producing rice for his family, as long as he continues to grow rice, Mr A. can enjoy the benefits of possessing and managing his own farmland with very little effort only on weekends. Even though it runs a deficit, rice farming is highly effective as a means to hold on to paddies passed down through generations, keep land use options open, and save on taxes.

Why are there so few young rice farmers?

On the topic of Japan's rice problem, the labor shortage or the ageing of rice farmers may be the first things to come to mind for some readers. It should be pointed out, however, that these are rice paddy issues, not rice problems. To begin with, there has been no such serious problem as the sudden disappearance of rice from the Japanese dining table at this point. Besides, Japan has been and continues to be completely self-sustaining as far as rice is concerned. While various commodity and service areas in Japan deplore their shortage of domestic labor, rice has enjoyed the most abundant labor supply among agricultural products in Japan. Seeing as rice is not in short supply, it is no wonder that new rice producers are hard to come by.

'Although we often hear on TV of the problem of a shortage of rice farmers, I've never felt that way myself,' said Mr B., a rice farmer from Tsuruoka, a city in Yamagata prefecture in the middle of the Shōnai plain, a rich rice-producing area. According to Mr B., as soon as a rice farmer gives up on farming due to old age, neighboring rice farmers scramble for use of the paddies.

Let us look closer at the issue of ageing. The average age of Japanese rice farmers is 67 years and the most common age group among rice farmers is over 70 years. The dearth of young farmers is clear. While 53% of new farmers grow vegetables, only 13% produce rice. Among those under 49 years old who have started farming, only as few as 10% chose to grow rice. On the other hand, one-quarter of new farmers over 60 years old became rice farmers. Surprisingly, it is the older farmers embarking on the second phase of their life who risk the overcrowded and fiercely competitive rice farming industry instead of the youth, who one thinks of as full of challenging spirit. One may wonder why.

Figure 1.6 shows a comparison of typical required hours of labor and expected income for growing rice and for growing asparagus and strawberries in Niigata prefecture, another rich rice-producing area in Japan. I invite readers to decide which crop they wish to grow, setting aside for the moment such details as land conditions

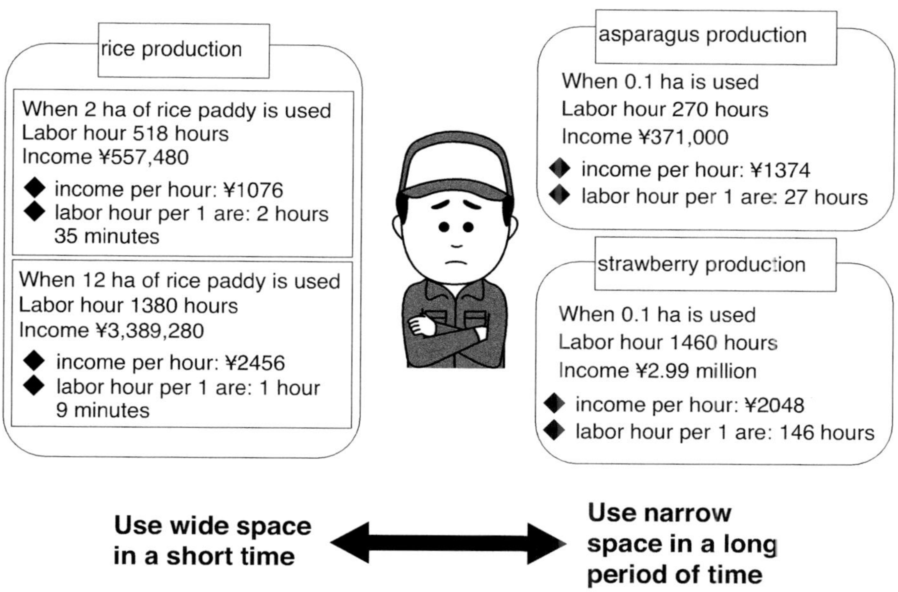

Fig. 1.6. Illustrated costs of rice, asparagus, and strawberry production. *Source*: based on the Ministry of Agriculture, Forestry and Fisheries' *Nogyo Keiei Tokei Chosa* [*Statistical Survey on Farm Management and Economy*] and documents compiled by Niigata prefecture. *Note*: income includes self-employment wages.

and the availability or absence of agricultural know how, facilities, machinery, and farm hands. Particularly, I'd like readers to pay close attention to the significant difference in the hours required, depending on the crop of choice. While 2 ha of rice paddies can provide only 518 work hours a year, 0.1 ha of farmland can promise 270 hours in the case of asparagus and 1460 hours in the case of strawberries in a year.

The hourly income from rice harvested in 2 ha of rice paddies is ¥1076. While the hourly income can be increased by expanding the scale of operation, one has to obtain more rice paddies for that. But it is not easy to rent or purchase farmland in Japan. Even if a rice farmer is more than eager to work extra hard, he will end up putting in labor out of proportion with his narrow span of farmland. And this is the reason young farmers tend to avoid rice farming. On the other hand, in terms of labor, vegetable growing is highly taxing work. With the same area, older people may prefer rice farming over growing vegetables because it is less burdensome. When a retired person inherits rice paddies from his or her parents or chooses to take up agriculture as a side job to supplement pension income, rice will be the easiest produce to grow, especially if the person also inherits rice farming machinery. These advantages enable a businessperson to engage in rice farming during weekends as a part-time farmer.

Rice paddies as a cause of weak international competitiveness

The difficulty in aggregating rice paddies is the main reason behind the weak international competitiveness of Japan's rice. If a farmland owner can combine neighboring tiny rice paddies into one large paddy or collect a number of paddies under their management with the managed paddies abutting each other, their scale of operation will be expanded, allowing for efficient use of machinery. Moreover, the aggregation of rice paddies has profound meaning for rice farming in Japan because doing so allows farmers to employ larger machines, which fundamentally changes rice farming methods. Altering rice farming methods will necessarily affect the cost of rice farming and, eventually, the sales price of rice.

The average area of a rice farm in Japan is about 1.4 ha, which is about as wide as the area of the Tokyo Dome baseball park in Tokyo. In contrast, the average area of a typical Californian rice producer is on a whole different scale: about 320 ha, which is about 70 times as large as the entire Tokyo Dome building. The typical tractor used in the United States has a 95–225 hp engine, which is much more powerful than the standard 20–50 hp tractor used in Japan. Because the machinery and farming methods employed in the United States (including seeding from an airplane) differ markedly from those in Japan, it only costs ¥200 to produce 5 kg of rice in the United States as opposed to ¥1283 for the same 5 kg in Japan.

The ideal and reality of scale expansion

Theoretically, the larger and the more concentrated an area of rice paddies (or the site of rice production) is, the less costly rice production becomes. In reality, however, rice paddies are often so scattered that combining them under one management does not lead to major cost reductions due to the time required for the farmer to move from one paddy to the next and to handle extra clerical procedures. While it has been confirmed that the introduction of big machines can cut the cost of rice production if a paddy is smaller than 150 ha, few farmers have efficiently collected enough paddies to warrant introduction of such machinery. The reality today is that the cost reduction effect declines as the paddy area exceeds 5 ha. Even when a farmer succeeds in collecting paddies in the hope of producing more rice, only 60% of the paddies owned is needed to produce the amount of rice allotted by the government to meet demand, leaving the question of what to do with the remaining 40%. Should the farmer wish to grow other crops, they will have to cope with the weed problem, which is not a serious issue for rice. And there is no guarantee that other crops will grow healthily in every rice paddy. Seeing as a rice paddy is tilled and prepared exclusively for rice, it is destined to be left unused, converted to other land uses including residential land, or abandoned (and thus wasted) when no suitable substitute crop is found.

Reality of rice paddy use and aggregation

Producing as much rice as one wishes or collecting only favorable paddies are in and of themselves not problems. In reality, however, many farmers have abstained from rice farming or knowingly collected paddies that are not too suitable for rice farming. The former of these two choices is called rice paddy reduction or rice production adjustment. This will be discussed in detail in Chapters 3 and beyond. Here, let me explain the latter choice.

It is not only aspiring newcomers to agriculture who have a hard time renting or purchasing favorable farmland. One reason is that, unlike other general commodities, the trading of farmland involves social relations. As an example, let's consider the case of Mr C. of Tsuchiura, Ibaraki prefecture. Mr C., who owns only 2 ha of rice paddies, decided to expand the scale of his operations by renting rice paddies. He now farms 30 ha of rice paddies and 1.2 ha of cultivated fields. Of the 63 former landowners who lent or sold rice paddies to Mr C., 54 were his friends or acquaintances, five were nearby neighbors, three were those introduced to him by friends and relatives, and one was a remote relative.[1] This shows that Mr C. relied on friends and acquaintances, blood relations, and people in the local community to acquire paddies.

Because obtaining a rice paddy is bound to involve these social and personal relations, there is no guarantee that farmers can collect only paddies that are favorable for rice farming. In some cases, farmers may find themselves in a situation where they have to agree to rent a rice paddy of inferior quality – from, for instance, an elderly landowner. In such a situation, farmers are more motivated not to betray the trust of someone close to them or to preserve a local rice paddy than they are to think only about their own benefit. It is not rare for the landowner of a favorable rice paddy to lend his paddy on the condition that an unfavorable paddy is part of the package.

What is an unfavorable rice paddy?

Some farmers may turn down another landowner's offer to use their farmland. The Ministry of Agriculture, Forestry and Fisheries, which looked into this issue, finds this happens most often when the farmland lot is too small or in poor condition. The second reason is when a farm is too remote (see Fig. 1.7).

Particularly noteworthy is the third reason for the offer to be declined, when the farm in

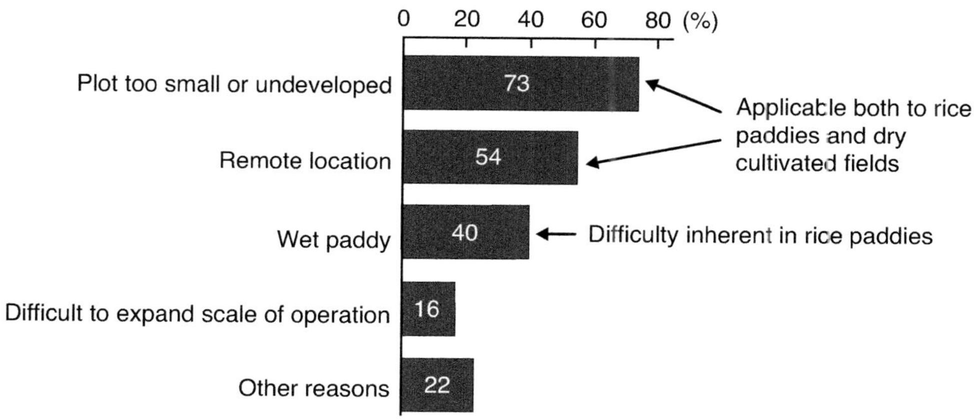

Fig. 1.7. Reasons farmers decline landowner requests to use farmland. *Source*: based on Fig. 2.2 of the Ministry of Agriculture, Forestry and Fisheries' *2013-nendo Shokuryō, Nōgyō, Nōson Hakusho* [*White Paper on Food, Agriculture and Rural Areas for FY2013*] (multiple answers from 450 corporations in 36 prefectures).

question is wet paddies unsuitable for upland crops. Now that rice paddies are in excess supply across Japan to the extent that rice is in excess supply, whether a paddy can grow other, non-rice produce such as upland crops seriously affects the marketability and rentability of the rice paddy.

Given our observations so far, we can now explain why in farm management centered around rice paddies, the larger the farmland area becomes, the lower the proportion of rice farming becomes, as shown in Figs 1.4 and 1.5. Even when farmers succeed in acquiring a large number of rice paddies, they will need to grow something else other than rice because rice is in excess supply. And it is easier to obtain paddies that can accommodate non-rice crops. As for farmers who produce rice for family consumption, their scale of business tends to be small. When all they can collect are wet paddies suitable for growing nothing else than rice, it doesn't make sense for them to expand their operations.

Rice paddy superiority/inferiority and problems particular to low uplands

Because each rice paddy is unique in the amount of sunshine it receives, its shape, its size, the width of the abutting farm road, and the timing and cost of water use, we can say that no paddy is identical to another. Even when a famer grows the same species of rice, uses the same fertilizer and agrichemicals, and gives the same care to each paddy, the quantity and taste of resultant rice differs from paddy to paddy. Some rice paddies are superior or inferior to the one next to them, and there is no guarantee that a particular rice paddy is a better one and will remain so in years to come.

Let us take the shape and size of a rice paddy, for instance. What decides the shape and size of a rice paddy? While natural conditions may come to mind first, technical standards may be an additional determinant. In contrast to a dry paddy for which having a slight slope does not matter, the surface of a water-filled rice paddy must be level. If a paddy is not level, the depth of water will differ for each rice plant, making uniform care impossible and resulting in growth differences between the plants in the paddy. The narrower the surface of a paddy, the easier it is to level out. We can easily imagine that if one tried to level the surface by hand, relying only on one's senses, it would be easier to do so in a hand-held bucket or pail than in a sizable paddy. This is one reason why rice paddies of old were small. As technology advanced to enable farmers to level out wider spaces for water, farmers became able to manage larger rice paddies, and thus use larger machinery. While the technology to create a level surface for water is merely one example, it goes to show that the advancement of relevant technology can drastically change the size and usage of rice paddies.

In parallel with the mechanization of rice farming, land readjustment vis-á-vis rice paddies also occurred on the national level. If rice farming were completely manual work, then the shape of a rice paddy would not affect work efficiency very much; be it a circle, a triangle, or a rectangle. When a machine is used, however, a rectangle is the ideal shape because it minimizes wasteful movements on the part of the farmer. The larger rectangle a rice paddy becomes, the more cost-efficient the farming can be due to economy of scale. In addition, the space of the ridge surrounding a rectangular rice paddy is minimized, thereby reducing the labor required for weeding.

Yet some rice paddies cannot keep pace with the advancement of technology. The most representative are rice terraces or, more broadly, rice paddies on low uplands; that is, the intermediary area between plains and mountains. Compared to paddies on plains, it takes longer working hours and more difficult labor to expand rice paddies on low uplands. Also, owing to the difference in levels between the surface of the rice paddy and its surrounding ridge caused by the slope, rice paddies on low uplands require a longer time to move around and extra care to weed. For rice paddies located on low uplands, the disparity in productivity only grows as the level of rice farming technology advances.

Although many rice farmers in low uplands produce brand-name rice, they still have to compete with rice growers on the plain who enjoy high production efficiency because their rice as a commodity is simply categorized as rice. This fact has bearing on the fatigue felt by low upland communities where rice production occupies the central position in industry. While rice growing in rice terraces and low uplands is often talked about only in terms of the tastiness of the products, the essence of the problems facing low upland farmers lies in the characteristics of their rice paddies. These characteristics, dictated by nature, cannot be salvaged by the tastiness of the rice.

A rice paddy with a wide slope, which will be very hard to work efficiently, meaning that it will be liable to become abandoned farmland in the future, like the plot towards the back of the photo. (Photographer: Masayuki Ogawa)

Abandoned arable land and a shortage of workers: are these rice problems?

Another aspect of the so-called rice problem that may come to mind for some readers is the increase in abandoned arable land. This is not a rice problem, however. An increase in abandoned arable land does not necessarily cause a shortage of rice. The situation is a rice paddy or farmland problem rather than a rice problem. Also, in the context of rice, a shortage of workers refers to a shortage of farmers using rice paddies. It does not point to a shortage of labor to actually grow rice.

Sometimes the increase in abandoned arable land and the dearth of rice farmers are seen as problematic phenomena related to rice growing. This is because the issue of rice paddies and farmland is mixed up with the issue of rice. The scarcity of lucrative farmers is one thing, but a scarcity of farmers who use rice paddies is a totally different story. Even if the number of farmers who profit from growing highly profitable crops in narrow farmland areas increases, this will not necessarily lead to an increase in the number of rice paddies in use.

Why is rice the Japanese staple?

When discussing rice as the staple food for Japanese, the essence of the issue cannot be captured unless its relations with rice paddies and national land use are taken into consideration.

For instance, some argue that rice farming must be cherished because rice is the staple food for Japanese. This argument, however, is a jump in logic caused by failure to separate rice from rice paddies. In this argument, eating rice is directly linked with producing rice. If rice is truly so important for the Japanese people, it would not be absurd to argue for the expansion of rice imports so that everyone in Japan can eat inexpensive rice.

Still, this misleading argument is voiced without hesitation. Why? Because the discussion about rice tacitly includes such non-rice issues as rice paddies and the national land. The expression 'rice as staple food' often includes rice's connection with rice paddies and national land use.

Some argue that it is erroneous to claim that rice is the staple food for Japanese seeing as the Japanese people had relied more on millet than rice as their staple food until the Shōwa period (1926–1989). Others refute this by saying that rice has long been the staple food for Japanese since ancient times. These two sides argue on different planes because, when judging whether rice is or is not staple food, some people take particular note of the physical quantity of rice while others consider other factors.

One of the earlier explorers of the reasons rice has been the staple food for Japanese is Tsunoda Shigesaburō, a Taishō-Shōwa plant breeder and geneticist. Paying special attention to plants' photosynthetic mechanisms and plant body structure, Tsunoda found that the balance between plant physiology and Japan's natural features is optimum in gramineous crops (including rice, wheat, and corn) and, thus, concluded that rice has been the staple food in Japan. In other words, Tsunoda defines rice as the staple food for Japanese based on Japan's

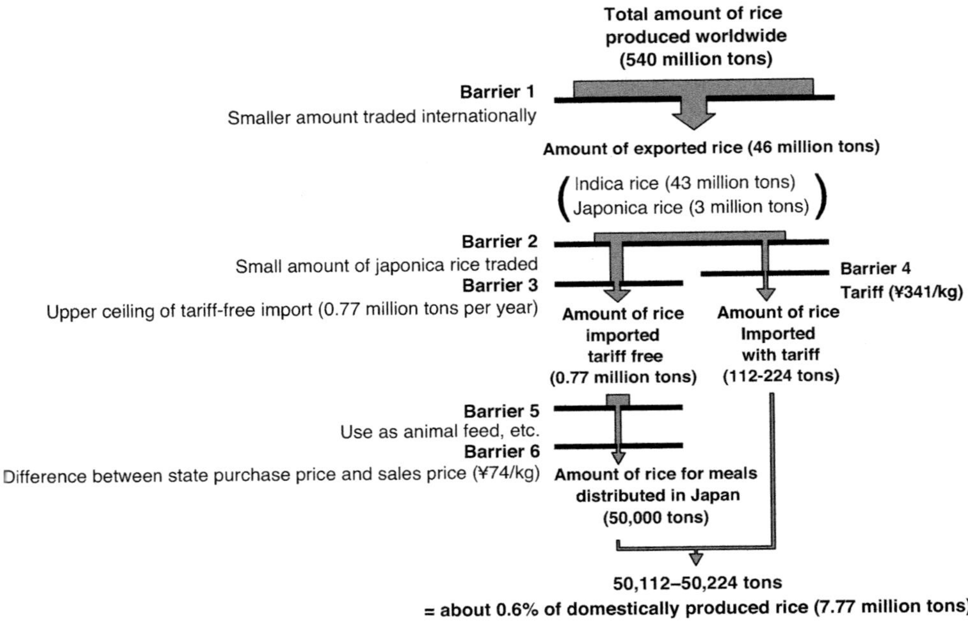

Fig. 1.8. Barriers to rice imports, 2020. *Source*: based on the Ministry of Agriculture, Forestry and Fisheries' *Kome wo Meguru Jōkyō ni Tsuite* [*On Situations Surrounding Rice*] and *Shokuryō Anzenhoshō Geppō* [*Monthly Report on Food Security*].

natural features, not the amount of rice that is eaten.

Barriers to rice imports

It is time to think once again whether rice is the staple food for Japanese just because it is eaten in large quantity or because it is rooted in rice paddies and Japan's national land. If international trade of agricultural products is further expanded, the time may come when Japan becomes a society that eats rice which it does not produce.

Figure. 1.8 identifies six import barriers for rice. Let me first point out that rice is unique in terms of the smaller amount it is traded internationally in comparison with other crops. Roughly speaking, there are more than 115,000 species of rice in the world, each with totally different tastes and flavors. The long and thin Indica rice is the mainstream species. Only a small amount of Japonica rice, the Japanese people's favorite, is being traded. As a matter of fact, this Japanese relish for Japonica rice has been a great import barrier.

Two kinds of rice are imported to Japan: tariffed rice and rice imported tariff-free by the Japanese government within the import ceiling. The tariff for rice is set higher than the price at which Japanese rice farmers sell their products in the domestic market. This puts imports at the extremely small amount of 112–224 tons annually. Moreover, rice imported tariff-free by the government is mostly used for non-food purposes, including animal feed, and is sold at higher than the purchase price. About 50,000 tons of rice is imported annually, bypassing the six barriers seen in Fig. 1.8, which comprises merely 0.6% of the rice produced domestically.

Of the six import barriers for rice in Fig. 1.8, numbers three to six are imposed by the Japanese government. They are both a wall to protect Japanese rice and a barrier to hinder Japanese consumers from obtaining overseas rice. It is difficult to judge whether these walls against importation are desirable. It becomes easier to judge, however, if one pays more attention to rice paddies instead of rice per se. Import barriers against rice are walls erected to maintain the area of rice paddies in use and prevent further increase in the surplus of rice paddies.

Developing Only the Rice Industry is Not the Solution

Rice industry: large margin for further growth

As we've seen, what is called the rice problem today is mostly a rice paddy issue, and so developing the rice industry will not automatically solve the rice problem.

Singled out, the rice industry still has a significantly large margin for further growth. Vigorous research and development as well as actual application have been carried out on robot tractors, rice paddy water management systems operated by smartphone, the use of drones, data sensing by satellite, and AI analysis of weather data, all for the sake of more efficient rice production. A new system has also been introduced, making it easier to rent and lend rice paddies between strangers.

Of all distribution channels for rice in Japan, supermarkets occupy about 50%, followed by a variety of channels such as direct marketing by farmers and internet marketing (15%); free gifts from family members, friends, and acquaintances (15%); and all other channels, including rice dealers (20%). Only two-thirds of the rice obtained through these channels is cooked and eaten by consumers. The remaining one-third is eaten in eateries or in the form of home-meal replacements such as frozen meals, boxed lunches, and rice balls purchased from convenience stores. As the patterns of distribution and consumption of rice diversify, so have direct marketing by rice farmers as well as manufacturing and marketing of pre-packaged cooked rice by companies also blossomed.

In addition, exports of rice and rice-based products such as rice crackers, sake, pre-packaged cooked rice, and rice noodles have been on the rise. The so-called gluten-free fad, which refers to avoiding gluten contained in wheat, has acted as a tailwind for rice exports, because rice is gluten-free. Exports of rice and rice-based products from Japan increased by 7900 tons or ¥8.3 billion between 2017 and 2020.

Less domestically produced rice is not a bad thing

The amount of rice consumed in Japan has been increasingly dwindling. The 7900 ton increase in rice exports is only 2% of the decline in domestic consumption of rice (360,000 t; see Fig. 1.9). Given this situation, some readers may worry about a decrease in domestically produced rice.

It has to be stressed here, however, that a decrease in domestically produced rice is not necessarily a bad thing. As the number of people who eat rice lessens, competition among rice brands becomes all the more heated. As a result, the more popular brands of rice and the more efficient methods of rice growing will prevail. When rice is over produced, its price will go down. This is all good news for people whose only contact with rice is as its consumers. Usage of rice other than as cooked rice is being explored, ranging from the development of rice suitable for business uses to research on new processing methods. It shows that room for development of the rice industry is wide open.

As domestically produced rice decreases, Japanese manufacturers of rice-related machinery are focusing their efforts on marketing their products overseas, utilizing the technologies they have nurtured in the Japanese market. Some young rice farmers have already given up on rice growing in Japan and started looking into overseas rice farming.

Shimmei Holdings is one of the largest rice wholesalers in Japan; it handles not only Japanese rice but also overseas rice. In addition to its overseas headquarters in the United States, China, Hong Kong, and Vietnam, Shimmei in 2021 founded a new company in Thailand, one of the world's leading rice exporters, through which it wishes to expand rice exports to Malaysia, the Philippines, and European countries.

It is not justifiable to speak of the rise and fall of rice farming in Japan solely on the basis of the rise and fall of the rice industry. The continued decline of rice consumption in Japan can

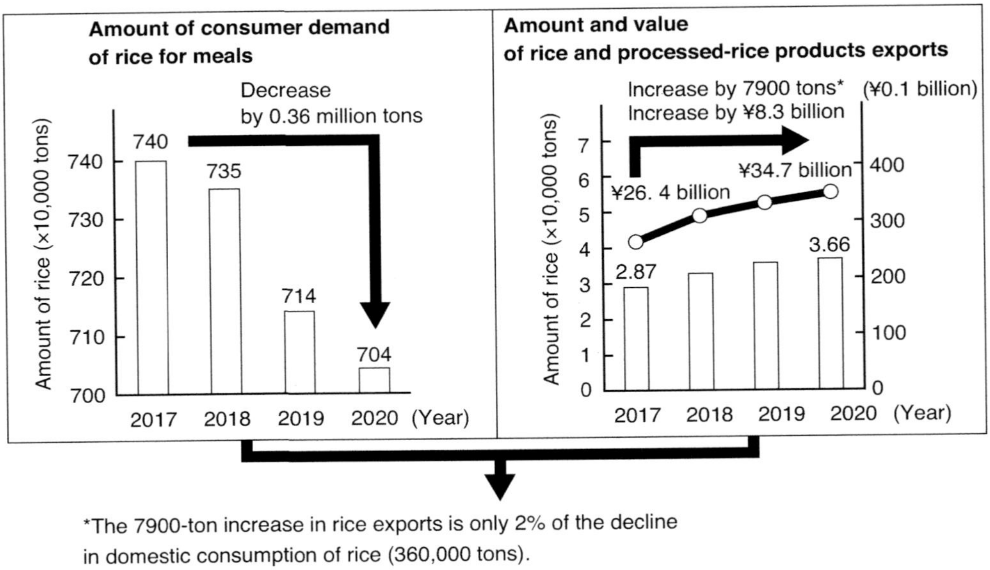

Fig. 1.9. Consumer demand for rice in Japan and export of rice and processed rice products. *Source*: based on the Ministry of Agriculture, Forestry and Fisheries' *Kome wo Meguru Jōkyō ni Tsuite* [*On Situations Surrounding Rice*] and *Kome & Kome Kakōhin no Yushutu Jisseki* [*Record of Rice and Rice Processed Product Export*]. *Note*: amounts of exports of processed rice products are in raw rice equivalent.

provide momentum for Japan's rice industry to flourish.

Rice paddies not growing rice is more serious than detachment from rice

In other words, the Japanese people's detachment from rice is not such a grave issue as far as rice itself is concerned. What is genuinely serious, instead, is the increasing number of rice paddies where rice is not produced. Using 1969 as the base year, when area for rice paddies as well as active rice paddies was the largest in the history of Japan, Fig. 1.10 shows changes in the amount of rice for meals harvested in Japan, the area of rice paddies, and the area of paddies used to grow rice for meals in the 50 years before and after the base year.

In the 50 years between 1919 and 1969, rice harvested for meals increased 1.5 times, while rice paddy area remained mostly unchanged. Also, the pattern of almost all rice paddies being used to grow rice remained essentially unchanged.

In contrast, in the 50 years between 1969 and 2019, rice harvested for meals decreased almost by half to close to the level of 1919, 100 years earlier. This does not necessarily mean that the situation surrounding rice production in 2019 was similar to that of 100 years earlier. The greatest difference between the two situations is the large decrease in total rice paddy area and the area of paddies producing rice for meals. Particularly worrisome is the reduction in the area of rice paddies where rice for meals was harvested. The rice paddy's shift away from rice is much more alarming than the Japanese consumers' shift away from rice.

Is it all right to be indifferent to the rice problem?

Once upon a time, rice was, both consciously and unconsciously, the center of talk in Japanese

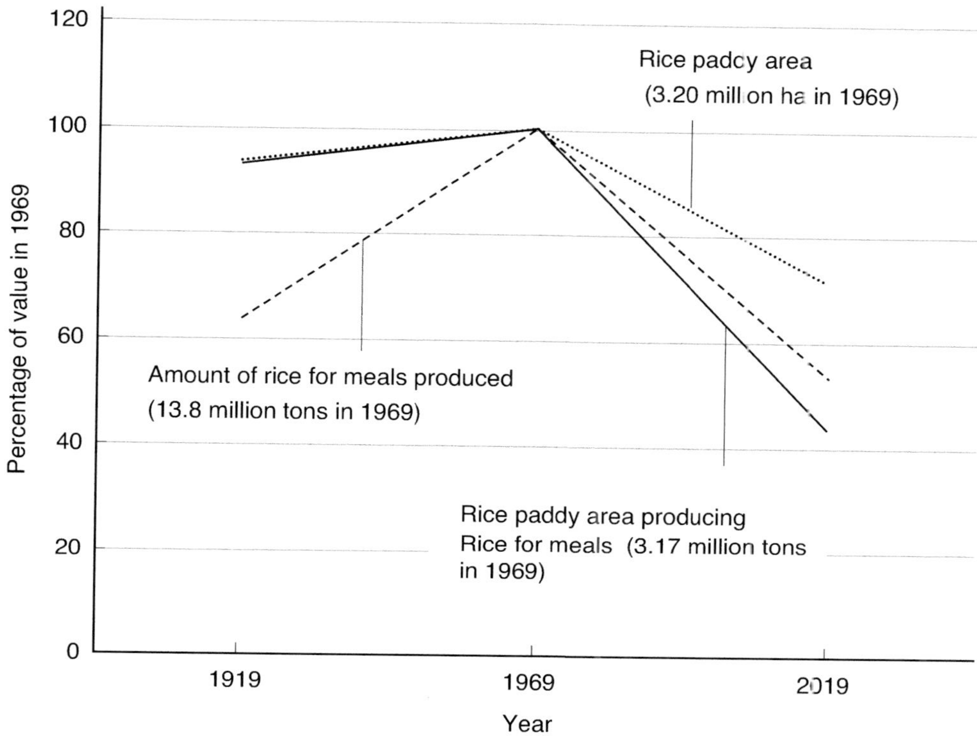

Fig. 1.10. Changes in rice paddy area, amount of rice for meals produced, and paddy area producing rice for meals. *Source*: based on the Ministry of Agriculture, Forestry and Fisheries' *Sakumotsu Tōkei* [*Statistical Survey on Crops*]. *Note*: all figures are index numbers for the entire Japan using 1969 as the base year.

society. As referred to in the ancient *Protocols of Engi Era Ritual Prayers* of *c*.900, where we find the words 'this year, in the second month, just as grain cultivation is about to begin,' annual rice growing has been perceived as an integral part of the flow of time. We see this exemplified by such annual rituals as seasonal festivals centered around rice and by times of the year being referred to as, for example, the busy farming season or the agricultural off-season. Regardless of one's social status or personal wealth, for people in ancient Japan, rice was a central concern. They shared celebrations of the harvest, aligned their schedules with the climate (which was beyond their control), and lamented bad harvests and famine. The prime example is the *Kokudaka* tax system based on rice during the Edo period, when rice rose to the status of a socioeconomic measure of Japanese society.

In contrast, today, when the share of farm households in total households nationwide declined to 3% and the share of agriculture in total gross domestic product (GDP) fell to 0.8%, it has become only farmers and those with a vested interest in agriculture who talk about agriculture, particularly rice. No wonder only a few Japanese are concerned about the issue of rice paddies.

As of 2022, Japan needs 1.26 million ha of rice paddies to produce sufficient amounts of rice. This is the equivalent of Hokkaidō and prefectures in Tōhoku and the Kantō-Kōshin areas plus Niigata prefecture in Hokuriku combined. Theoretically, if rice is produced in those northern regions, no prefecture west of Fossa Magna, a great rift lowland in the center of Honshū island, needs to grow rice. In addition, rice consumption in Japan has decreased

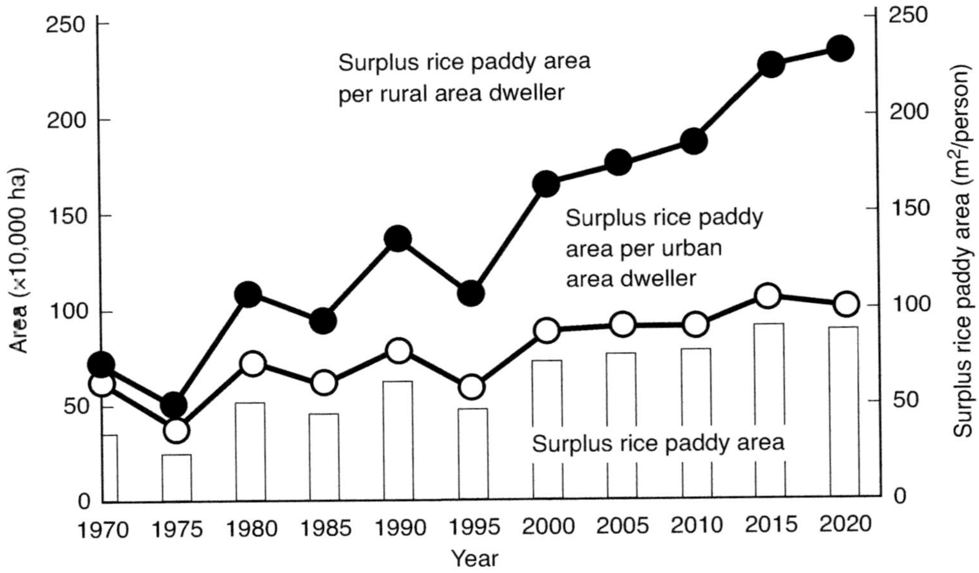

Fig. 1.11. Changes in surplus rice paddy area. *Source*: based on the Ministry of Agriculture, Forestry and Fisheries' *Sakumotsu Tōkei Chōsa* [*Statistical Survey on Crops*] and the Ministry of Internal Affairs and Communications' *Kokusei Chōsa* [*Population Census*]. *Note:* an urban area is an area with high population concentration. A rural area is an area with low population centration.

by 100,000 tons yearly, which is the amount of rice that Gifu prefecture alone produced in 2021. Even if this declining trend continues, it is inconceivable that Japanese meals will become totally devoid of rice. But we are right to fear that rice farming and the Japanese people will grow increasingly apart. In the more than 3000 years of its history, rice production in Japan has never faced such a crisis situation before.

There is today some 0.9 million ha of surplus rice paddies in Japan, or the equivalent to the total area of Kagoshima prefecture (see Fig. 1.11). Calculated based on urban dwellers, the size of the rice paddy surplus per capita has not greatly changed in the past half century. But surplus rice paddy area per agrarian person grew to 233 m^2 in 2020: the size of a tennis court.

It should be pointed out that the perception gap between urban areas and agrarian villages on the fundamental rice problem, including rice paddy issues, will continue to widen year by year. Moreover, as the vote/value disparity is reduced on the basis of population under Japan's parliamentary democracy, agrarian villages' problems of dwindling population and a growing surplus of rice paddy area could become a source of social division unless efforts are made to explore new policies and educate urban dwellers to gain their understanding.

The issue of surplus rice paddy area is more than a contemporary problem. It is an important theme that bridges the past and the future; that is, how the Japanese people should or should not pass on the assets they have inherited from their forebears to future generations.

Some readers may think the rice problem is an old, worn-out issue. If, however, rice paddies are perceived to be an integral part of the rice problem, then it is a problem about which every Japanese needs to understand the essence. What is called for is not merely knowledge on rice production, distribution, and consumption but also insight into the depth of the problem, including the rice paddy issue that is at the heart of the rice problem. The issue of rice and rice paddies, despite their indelible presence in the history of the Japanese people, appears to have been deprived of the attention it used to enjoy. Paying closer attention to the rice problem once again may lead to insights into the greatest and imminent crisis Japanese society is facing as well as into the choices that the Japanese can take vis-à-vis this crisis today and tomorrow.

With this in mind, let us look into the four historical turning points so crucial to an understanding of today's rice problem, as well as a fifth turning point that the Japanese people are bound to face.

Note

[1] Based on Isaka Tomomi. 'Nōchi torihiki mekanizumu no shoruikei to hishijōteki torihiki no jittai' [Classification of Farmland Transaction Mechanisms and the Actual Situation of Non-market Transactions], *Nōgyō Keizai Kenkyū*. Vol. 89 No. 1, 2017. The total of 63 included one landowner whose relations with Mr C. is unknown.

2 Self-Sufficiency in Rice Attained: The First Turning Point in 1967

Abstract

Chapter 2 focuses on the year 1967, when Japan attained self-sufficiency in rice for the first time. That year is defined as the first turning point since the beginning of rice production in Japan. Analyzing situations surrounding rice from the Meiji era to immediately after World War II, this chapter details how 1967 became a major dividing ridge between the days of rice production to evade famine and the era of rice cultivation liberated from fear of starvation. The chapter also finds it problematic that Japan's rice production was devoid of any vision beyond attainment of rice self-sufficiency after the first turning point.

The Ardent Wish to Accomplish Rice Self-sufficiency

Rice and population

It was in 1967 that Japan began to produce more than enough rice to feed its population. That year marked a turning point after which the history of rice and the Japanese people changed completely, as if they were a totally new pairing.

Until this turning point, the relationship between rice and the Japanese people, which had spanned more than 3000 years of history, had seen the parallel growth of rice production and population. Started in the Jōmon period (c.14,000 to 1,000 BCE), rice farming spread to all corners of Japan, except Hokkaidō, during the Yayoi period (10th to 3rd centuries BCE). Rice farming became a stable source of food supply, much more so than hunting, food gathering, or slash-and-burn farming. This led to population growth. With a rise in the number of workhands, land was cleared and new rice paddies were cultivated one after another, contributing to the further spread of rice farming.

Rice paddies first appeared in western Japan on mountains and the watersheds of streams, where it was easy to control water and contain flood damage. Advancements in civil engineering and water management technologies eventually allowed the cultivation of land along bigger rivers and the conversion of wide swamps to rice paddies. As a result, an increasing number of rice paddies were created on plains.

The aggressive development of new rice paddies in the early Edo period (1603–1868) was particularly impressive. Rice paddies mushroomed all over Japan, including in the Kantō plain, where swamps were converted to rice paddies. Total rice paddy area nationwide more than tripled in the mid-Edo period from 950,000 ha in the mid-Muromachi period (1336–1573). The population of Japan also more than doubled in the same period. While population growth stalled or even declined occasionally due to delays in new paddy development or bad harvests, over the long run both rice farming and Japan's population continued to grow steadily together. In particular, the population began to grow gradually from the end of the Edo period, a period relatively free of bad harvests or epidemics. The Meiji Restoration of 1868 triggered Japan's industrial revolution. As industrialization and urbanization were promoted, population growth maintained its momentum. In order to fill the stomachs of a growing population, rice farming was actively promoted. In the end of the Edo to early Meiji periods, the population growth rate superseded that for the first half of the Edo period. Japan's population surpassed the benchmark of 100 million in 1967, 100 years after the first year of the Meiji period (1868–1912). This means that in 100 years the population tripled from 34 million in the early Meiji period—the most rapid population growth in Japan's history.

DOI: 10.1079/9781800627093.0002

It should be noted that in 1967 Japan achieved self-sufficiency in rice. This shows that the history of Japan until its population reached 100 million was that of the Japanese people's quest for a society where they could eat domestically produced rice to their heart's content.

Rice became the staple of the Japanese people

While the growth in rice production and the population occurred in tandem, the Japanese people did not eat only rice. In fact, in some areas, people attached more importance to cereals and potatoes than rice, while people often mixed vegetables and other ingredients with rice when they ate it. To begin with, under the feudal socio-political system, many rice farmers in those days could hardly eat the rice they produced.

> Japan is a nice land of nice paddies and nice rice
>
> I harvest this nice rice on this nice paddy but I can never taste it myself
>
> Even though I harvest rice, my ten children can only sip porridge
>
> Even though we tend cocoons to weave silk thread, we can only wear rags

These are two examples of *dodoitsu*, a form of Japanese poetry developed towards the end of the Edo period, depicting the hard life of peasants in the Edo period (Takahashi, 1977). Peasants were forced to pay two-thirds of their harvests to feudal lords (*daimyo*) during the rule of Toyotomi Hideyoshi in the late 16th century and one-third to the shogun during the Edo period. This was a tremendous burden for peasants. Although the rule of the samurai ended with the Meiji Restoration, exploitive relations between landlords and tenant peasants remained unchanged in that peasants continued to submit more than half of their harvest to landlords.

Triggered by the adoption of a conscription system and food rationing in the Meiji period, however, rice came to be eaten routinely all over Japan by everyone. If something that is eaten frequently deserves to be called the staple food, then rice attained that position with war as its backdrop. As rice came to be regarded as military goods, the amount of rice that was produced and eaten in Japan began to increase rapidly.

While regional disparities in the proportion of rice as the staple food existed – that is, the proportion was larger in rice paddy-rich regions and lower in areas with few rice paddies – these gradually evened out along with the development of transportation systems. During the 10 years after the institution of the 1873 Conscription Ordinance, a total of 3.75 million tons of rice for meals was produced nationwide, including rice that farmers secretly harvested without the government's knowledge. Because the Japanese in those days did not eat as much wheat or meat as today, rice alone could not fill their energy needs. Nevertheless, if these 3.75 million tons of rice were distributed evenly among 37.62 million people, the total population of Japan at that time, each person would have eaten about twice as much rice as today.

Taiwan and Korea: rice depots for the Empire of Japan

It should be noted that the amount of rice consumed and the amount harvested did not necessarily grow smoothly at the same pace. In the years between 1875 and 1897, both the amount of rice eaten per capita and Japan's population increased 1.2 times each, pushing the total amount of rice eaten to 1.44 times more (1.2 × 1.2). In contrast, rice harvested in the same period grew 1.39 times. The harvest could not keep up with the rise of both per-capita consumption and the population. As the gap between the two widened year by year, rice imports increased to make up for the deficit.

In 1897, Japan's rice imports exceeded exports by 193,000 tons. Over the next 70 years until 1967, the Japan mainland experienced rice shortages as a matter of course. To escape the constant rice shortages the Japanese government instituted the policy of promoting rice production in Taiwan and Korea, which Japan had colonized or annexed in the late 19th to early 20th century, and shipping part of the harvests to the Japan mainland.

In Taiwan, plant thremmatologist Iso Eikichi and his associates succeeded in breeding an improvement on the local rice that satisfied Japanese people's tastes. The new breed was called Hōraimai. Under the leadership of civil engineer Hatta Yōichi, the Usanto Dam, the largest dam in Asia in those days, was constructed. As a result, 150,000 ha of Taiwan wilderness were converted into rice paddies where Hōraimai, sugar cane, and vegetables were grown. In Korea as well, rice production also increased greatly: rice paddy area expanded, the use of fertilizer escalated, and newly improved rice breeds were introduced.

In the 30 years between 1909 and 1938, rice harvest yields increased by 2.1 times in Taiwan and 3.2 times in Korea. The amount of rice shipped to Japan from these overseas territories rivaled inter-prefectural rice distribution in Japan, making it an essential asset to the Empire of Japan. Moreover, in fact, rice imports from these areas grew so much that some argued that Japan should control domestic rice production.

Rice shortage leads to acceptance of Potsdam Declaration

The scheme of importing rice from Taiwan and Korea, however, was by no means without flaw. Due to shortages of labor and fertilizer as well as severe drought in Korea, the amount of rice transported to Japan declined sharply.

The most critical weak point in the scheme was maritime transport. At an Imperial Council meeting convened a week before the start of the war with the United States in 1941, Emperor Shōwa asked about food supply prospects. Ino Hiroya, minister for agriculture and forestry, replied that Japan would not suffer from food shortages for 3 years as long as the Imperial Navy secured vessels to transport the rice. The security of the maritime route used to transport rice from Taiwan and Korea was a matter of life or death for Japan. As a matter of fact, as the course of the war worsened, the sea route was blockaded by U.S. forces, stagnating rice transport from Taiwan and Korea. The situation was so dire that, just before the end of the war, the Inspectorate General of Shipping Operations, Imperial Headquarters, issued an order for all vessels to prioritize embarkation of food and salt even over suicide attack weapons, ammunitions, and bombs, and aircraft fuel (Tanaka, 1974).

On August 9, 1945, when the Soviet Union declared war against Japan and the United States dropped the second atomic bomb, it was announced at the extraordinary cabinet meeting that Japan had to brace for a severely bad rice harvest. This news was the final impetus for the imperial decision to accept the Potsdam Declaration; the decision was issued the next morning before dawn. As Emperor Shōwa himself later reminisced during a televised interview with the National Broadcasting Company (NBC), an American television broadcaster, the shortage of food was one major reason behind his decision to end the war.

Postwar Japan starts with a severely bad harvest

The rice harvest of 1945 was merely 67% of that of the previous year. That year's record for worst harvest remains unbroken to this day. Among causes of the unprecedented bad harvest were, on top of an extraordinarily cold summer, a shortage of farmhands due to military conscription and an insufficient supply of fertilizer. Post-World War II Japan, deprived of Taiwan and Korea and hit with a massively bad harvest, took its first steps under a food shortage that was much more severe than any seen during wartime. Japan's population grew rapidly, including 5 million repatriates from overseas territories, but agrarian villages exhausted by the war simply did not have the productivity necessary to satisfy the food needs of the growing population.

In October 1945, it was announced that Tokyo's rice stockpile had dwindled to no more than a 3 day supply. Flabbergasted by this announcement, then Finance Minister Shibusawa Keizō made a remark which triggered the circulation both at home in Japan and abroad of a rumor that 10 million people would die of starvation.

On top of the shortage of rice, the rice rationing system did not function properly. Since 1940, rice farmers had been obliged to sell all of their rice harvest, excepting that for family consumption and seed planting, to the national government. Because the amount to be sold to the government was precalculated instead of being determined by the actual harvest, farmers ran short of rice to sell when struck by an exceptionally bad harvest. In such cases, rice farmers were allowed to substitute coarse cereal, potatoes, sweet potato stems, or even acorns for the amount of rice they were short. But seeing as rice farmers had to sell even rice and potatoes that had been intended for their own consumption, this system put a tremendous burden on farmers, particularly those in areas that could produce only rice.

Most of all, the low price paid by the government impoverished rice farmers. This became one of the causes of rice farmers selling rice on the black market. By October 1945, agricultural produce procured either legally or illegally was being sold on the approximately 17,000 black markets that had emerged all over Japan. Of all the commodities, milled rice, nicknamed 'silver rice,' was the prime item traded. At times, milled rice sold for 132 times the standard price sanctioned by the government. By early January 1946, the Japanese government had procured a total of 1.13 million tons of rice. Compared to the 3 million tons it had collected around the same time in the previous year, this was an astonishingly poor amount.

In an effort to break free of this situation, the General Headquarters (GHQ) of the Supreme Commander for the Allied Powers requested the Japanese government in February 1946 to use the power of GHQ and Japan's own law enforcement authorities to tighten the crackdown on farmers who refused to deliver rice as well as on black market rice dealers. Niwa Masajirō, an official at the Food Control Bureau, an affiliated agency of the Ministry of Agriculture and Forestry (the predecessor of the present-day Ministry of Agriculture, Forestry and Fisheries) wrote in his memoir that GHQ made the request as a way of demonstrating to the overseeing Far Eastern Commission that efforts were being made to acquire rice. Because it was the policy of the Far Eastern Commission to minimize food consumed by the Japanese people and curb food imports as much as possible, GHQ tried to obtain approval of the Commission for food imports by stressing it was doing everything it could to alleviate the food shortage.

Three months later, mass protests around the country, including a May Day demonstration in which 250,000 Japanese gathered outside the Imperial Palace demanding rice, reflected the fact that Japan had a food crisis. In response, Emperor Shōwa appealed for national solidarity through a radio address, his first since his surrender speech on August 15, 1945. Agriculture Minister Wada Hiroo also stressed in a radio address that, 'These times are far more important than the wartime.... We are truly on the verge of our nation's extinction or resurrection' (Wada Hiroo Ikō shū Kankōkai, 1981). General Douglas MacArthur, Supreme Commander for the Allied Powers, hurriedly promised Prime Minister Yoshida Shigeru that the United States would send additional food aid to Japan. The Japanese narrowly escaped mass starvation and death because of this aid. Nevertheless, Japan's chronic food shortage persisted until the mid-1950s.

During the time immediately following the war, a rice substitute became popular, partly thanks to promotion by the Japanese government. Made of imported wheat and other ingredients to approximate the shape and texture of rice, this rice substitute was a uniquely Japanese attempt to please those people who did not care for wheat and desperately wished to eat rice. One can detect here the Japanese people's tenacious attachment to rice. For the Japanese in those days, self-sufficiency in rice – meaning being able to fill themselves with as much rice as they wanted – was literally a fervent wish.

Goal Attainment Without a Sense of Achievement

The unknown turning point

At a cabinet meeting on August 29, 1967, Kuraishi Tadao, minister of agriculture and forestry, reported that the year's rice harvest was predicted to be an unprecedented bumper crop. In contrast to the smiling Kuraishi, Prime Minister Satō Eisaku wore a look of mixed emotion as he murmured to Kuraishi, 'Had Mizuta been here, he would have voiced his perplexity.' Satō was thinking of Finance Minister Mizuta Mikio, who was absent from the meeting that day (*Nihon Keizai Shimbun*, 1967).

No major daily newspaper, for that matter, saw the predicted bumper crop as great, welcome news. In fact, the news received with a muted reaction, considering there still were many Japanese suffering from the rice shortage and acute starvation. One has to wonder about the cause for such an unenthusiastic reaction to the accomplishment of self-sufficiency in rice, a longtime ardent wish of the Japanese. Next let us look into the reason for this indifference from multiple angles. Our inquiry should also present us with a hint as to why the Japanese people have continued to lose interest in rice.

Demonstrators marching in front of the Imperial Palace Plaza (*The Asahi Shimbun*, May 19, 1946) (Permission from from Asahi Shimbun)

Why 1967 is regarded as the year of self-sufficiency in rice

Prior to 1967, Japan did have a history of rice surpluses due to good harvests, and rice exports were a means for Japan to obtain foreign currency. While, admittedly, it is a bit of trivia, let me explain how the rice surpluses after 1967 were decisively different from those before that year.

An important point to keep in mind here is that, if self-sufficiency in rice is perceived to be achieved when the amount of rice produced equals the amount consumed instead of the amount needed, then self-sufficiency can be considered to have been attained even when people are starving from a shortage of rice. Prior to 1967, the amount of rice eaten had continued to grow in Japan as a whole as long as there was rice to be consumed. After 1967, however, the amount of rice eaten did not always grow even when there was rice to be consumed. This was a decisive departure from the previous pattern. Therefore, it must be said that 1967 represents the first major turning point in the history of rice in Japan since the beginning of rice farming in the Jōmon period.

To go into more detail, one's conclusion as to whether rice sufficiency was achieved will differ if one includes in the calculation rice used as an ingredient in such processed products as sake and miso as well as rice used as a seed plant or animal feed. Also, after comparing the amount of rice harvested and the amount consumed, some researchers argue that 1966, not 1967, was the year rice self-sufficiency was achieved. Still other scholars, myself included, designate 1967 as the year of rice self-sufficiency in the sense that that year all the rice needs in Japan were met with domestically produced rice. Some even call 1967 a year of full self-sufficiency in rice.

Rice self-sufficiency is a unique turning point in the sense that the fact that it was achieved was only realized in retrospect after quite some time. It was not possible for the Japanese in 1967 to judge whether rice self-sufficiency was attained because it was still possible that more rice could be consumed than predicted. And this is why people did not feel like rejoicing over the exceptionally good harvest of 1967 without reservation. While this historical accomplishment of rice self-sufficiency in 1967 is introduced in such relatively recent documents as the Ministry of Agriculture, Forestry and Fisheries' *Nenji Hōkoku 50-nen wo Furikaette* (*50 Years of Annual Reports in Retrospect*), an attachment to the ministry's 2010 White Paper, it is not even mentioned in older documents, including the Food Agency's 1980 publication *Nippon no Shokuryō Shigen – Nipponjin to Kome no Rekishi* (*Japan's Food Resources—History of the Japanese People and Rice*).

Good harvest: from gift of nature to fruit of human efforts

The Ritual Ceremonies of the Imperial Palace include the ancient Niinamesai (ceremonial offering to the deities by the emperor of newly harvested rice). As exemplified by this imperial tradition, the Japanese people have thanked heaven for good harvests every year since ancient times. Why, then, have the Japanese come to care less about good harvests? One possible explanation for the cool reaction to the news of a bumper crop in 1967 is a shift in the perception of a good harvest as being a gift of nature to being something that can be obtained through human effort.

Traditionally, a good harvest was usually explained in connection with the weather. All that rice farmers could do, under those circumstances, was to do their best and leave the rest to heaven. After analyzing 1967's good harvest and finding that the year's weather had not been exceptionally good compared to adjacent harvest years, however, the Ministry of Agriculture and Forestry concluded that it would have been a good harvest year anyway even if the weather had been just normal. In fact, in the summer of 1967, western Japan was experienced an unprecedented drought that gravely damaged the rice crop, while Yamagata and Niigata prefectures were ravaged by torrential rains (the Uetsu Flood) just before harvest season. Nevertheless, the rice harvest that year on the national level increased as much as 12% from the average crop.

One reason for the good harvest was the advancement of technology. The government's powerful promotion of increased food production to cope with the food shortage resulted in the advancement of relevant science and technology. As such, the rice harvest per hectare of paddy increased sharply from 3 tons in 1941, the year war with the United States began, to 4.5 tons in 1967.

Further momentum came from rice farmers' enthusiasm. Thanks to the land reform conducted immediately after World War II, rice became a possession of its harvester. Also, tenant rent was greatly reduced. From the viewpoint of tenant farmers, this meant that Japanese society was reborn to where their efforts in rice production were rewarded, making them much more motivated to grow rice. In a project to promote competition to become the no. 1 rice farmer sponsored by a major daily newspaper for the 20 years following 1949, as many as 20,000 farmers participated every year, some of whom succeeded in producing more than twice as much rice as the average.

Particularly noteworthy was Sagae Kin'ichi of the Yonezawa basin, Yamagata prefecture. Some 30,000 rice farmers visited Sagae every year to study his farming methods. Having inherited 2.2 ha of rice paddies from his father, Sagae studied rice farming diligently, including apprenticing in the fall of 1946 at the age of 24 with Tanaka Shosuke, an exemplary local farmer. In 1950, Sagae succeeded in harvesting 7.95 tons of rice per hectare. Although he abstained from participating in the newspaper-sponsored competition, his was a record that would have beaten the first prizewinner that year. Soon a research group was formed around Sagae, who persisted in his study of rice farming.

The northward advance of rice farming

During the Edo period, the value of a feudal territory was measured by its potential rice harvest. Only the territory of the Matsumae-*han* in the southern tip of Hokkaidō was at first registered as having zero potential. This is because it was believed that Hokkaidō was too cold to grow rice. It was only in 1692 that rice production was started in Hokkaidō. Over the subsequent 270 years, however, Hokkaidō perfected rice growing to the point that in 1961 the prefecture boasted the no. 1 rice harvest in the country.

The advance of rice farming to cold regions enabled by technological advancements also contributed greatly to the good harvest of 1967. For instance, a new method for raising seedlings called the protected semi-irrigated rice nursery, which was developed in 1947, rapidly spread among rice farmers because it allowed rice production even in areas where water is cold. Also, new breeds of rice resilient to colder weather were developed, among which the Nōrin 33-gō breed was proven to be technically viable throughout almost the entire expansion of Hokkaidō. In the similarly cold Tōhoku (north-eastern) region of Honshū, the main island of Japan, the spread of the Fujisaka 5-gō breed and other new breeds resilient to damage from the local cool summers contributed to an expansion in the areas where rice could be harvested (see Table 2.1). In the 1960s, rice harvest per hectare of paddy in Tōhoku and Hokuriku, another north-western part of Honshū, superseded that of western Japan, where, relatively speaking, rice farming had become widespread much earlier. This reversal phenomenon still lingers today.

Table 2.1. Rice paddy areas producing rice and the amount produced, in 1958 and 1967. Source: based on the Ministry of Agriculture, Forestry and Fisheries' *Sakumotsu Tōkei Chōsa* [*Statistical Survey on Crops*]. *Note*: the amount change was calculated before rounding.

	Area (10,000 ha)			Amount of crop (10,000 tons)		
	1958	1967	Percentage change	1958	1967	Percentage change
Hokkaidō	19	25	133%	72	111	156%
Tōhoku	58	63	110%	231	327	141%
Kantō-Kōshin	51	54	104%	184	226	123%
Hokuriku	36	36	100%	146	176	120%
Tōkai	28	26	93%	95	103	108%
Kinki	29	26	90%	109	110	101%
Chūgoku	30	29	95%	112	126	113%
Shikoku	15	14	95%	55	58	105%
Kyūshū	42	42	100%	166	188	113%

Increasing number of square rice paddies

The total area of rice paddies in Japan was 3,415,000 ha in 1967, an increase of 430,000 ha from 1945. Based on the average harvest per hectare in 1967, we can calculate that those 430,000 ha could produce 1.73 million tons of rice. This is exactly the total amount of healthy rice harvested and transported to the Japan homeland from Taiwan and Korea before the Pacific War. Having lost Taiwan and Korea after the war, Japan was finally able, some 20 years later, to make up for the rice it had lost of its own accord.

In addition, the Land Improvement Act enacted in 1949 prompted radical changes, such as the consolidation of the shape of rice paddies, water ditches, and farm roads. This, along with wider use of agricultural machinery, led to an increase in square-shaped rice paddies. Tractors that had been a focus of research and development since pre-war days not only helped reduce the burden on rice farmers in the paddies, they also liberated farmers from the time and labor to rear cows and horses for plowing. A particular type of tractor called a walking tractor rapidly grew popular among rice farmers after 1956.

The longing for rice and more rice paddies

The increase of rice paddy area in the 1950s and 1960s was mainly attributable not only to deforestation and land reclamation but also to the conversion of dry fields to rice paddies. For instance, in areas short of water for rice paddies, local farmers got together to set up land improvement districts suitable for cultivating rice. In some areas, former sericulture farmers for whom business had fallen off due to the rise of synthetic fabrics converted their mulberry plantation fields into rice paddies. The constantly arid Ojiyakegahara region of Niigata prefecture located on a fluvial terrace of the Shinano river used to be filled with mulberry plantation fields and miscellaneous trees. Anxious to harvest rice there, local farmers set up a land improvement district in 1968, 1 year after the first turning point of 1967. These farmers succeeded in drawing water from the Shinano river, which runs 140 m below the fields, by constructing an irrigation pumping station. As a result, by 1971 the area was filled with fertile rice paddies.

The dry rice paddy construction movement started in Mitakaya-mura (presently Hanyū-shi), Saitama prefecture, in around 1935 to convert fields, orchards, and woodlands to rice paddies. It spread rapidly throughout the northern Kantō region after the war. In addition, after 1956, vinyl rice paddies – where a vinyl sheet is placed underneath the soil to prevent water from leaking out – gained wide usage, particularly in sand dune areas. In 1967, as much as 200 ha of vinyl rice paddies were cultivated on the Shōnai Sand Dunes in Yamagata prefecture. At the time of the first turning point in 1967, new rice paddies in a number of places were in the middle of cultivation, and settlement in the reclaimed land of the Hachirōgata lake, Akita prefecture, had started in

January of that year. As the momentum for rice paddy cultivation and rice harvesting continued, Japan was about to enter an era of rice surplus.

In conversations with farmers in areas where dry-field farming was popular, I have often heard farmers born in the 1930s and 1940s telling of experiences in their youth of digging wells to use for rice paddies. Mr D. of Sagamihara-shi, Kanagawa prefecture, in the vicinity of Tokyo, who was born in 1937, dug a well with his mother and sister when he was 20 years old so that 24 ares of field could be cultivated for rice. Many of Mr D.'s neighbors also rushed to dig wells. In the end, the paddies of five neighboring farmers expanded to 120 ares. Mr E., who was born in 1940 on a terrace in Katori-shi, Chiba prefecture, where potatoes were the staple instead of rice due to poor irrigation, vividly remembered digging wells with his wife. He nostalgically reminisced, 'Driven by a longing for rice, which has been with me since my childhood, I cultivated a rice paddy in my hometown where I grew up.'

(Top) A cliff being excavated to construct the Ikegahara irrigation pumping station (photo taken at the land improvement district in the southwest of Ojiya) (Photograph used with permission). (Bottom) Present-day Ikegahara district, Ojiya-shi, Niigata prefecture. (Photograph by Masayuki Ogawa)

Why farmers lacked a sense of achievement

As the result of technological innovations on a number of fronts, actions taken by enthusiastic farmers, and the increased area of rice paddies, the serious rice shortage problem was resolved in 1955, only 2 years after the processed rice made of wheat and other ingredients entered the limelight.

When more than 80% of farmers were harvesting rice, it was the farmers who were most anxious to accomplish rice self-sufficiency. As it became increasingly easier to harvest rice, however, the number of farmers themselves began to dwindle. Although the percentage of those engaged in agriculture in the entire working population was over 45% in 1950, it began to decrease rapidly thereafter. In 1967, it dropped below 20% for the first time.

In around the first turning point year of 1967 was the period when new breed improvement methods, including radiation-induced mutation and anther culture, were put to practical use and such new machines as the riding tractor, rice-planting machine, seeding machine, and combine harvester were commercially marketed. The rice-planting machine used in this period was the type that a farmer pushes while walking and which plants a row or two of rice at a time. It was after 1975 that the riding rice-planting machine, which plants more than four rows at once, became widely used. By the 1960s, Japanese manufacturers had acquired the ability to develop agricultural chemicals themselves, and this prompted the full-fledged promulgation of pesticides, fungicides, and chemical herbicides specifically developed for rice farming.

Twelve years after the serious food shortage problem was resolved and in the midst of circumstances that remained favorable to rice farming for some time to come, Japan's having achieved rice self-sufficiency in 1967 had already degenerated from the attainment of an ardent wish into a mere pass point en route to more ideal methods of rice farming.

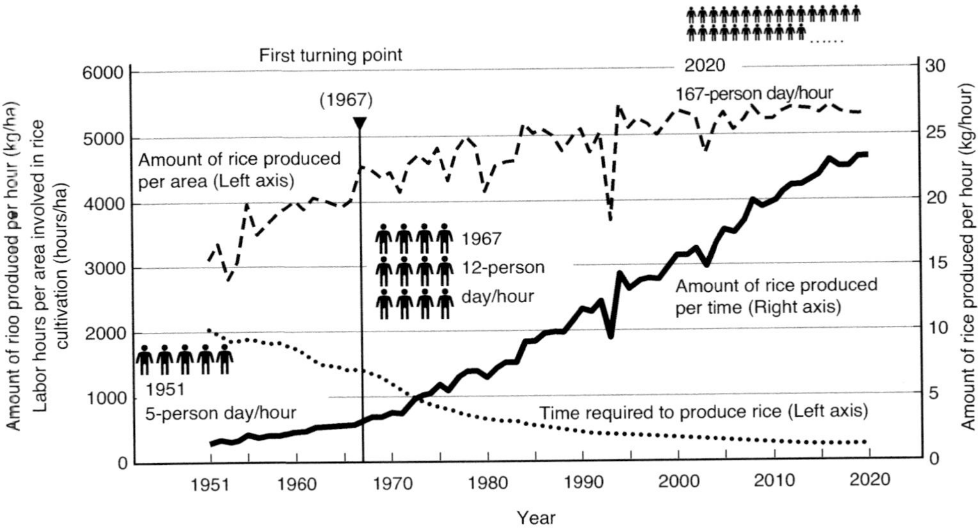

Fig. 2.1. Changes in the time required to produce rice, amount of crop, and amount produced per hour. *Source*: based on the Ministry of Agriculture, Forestry and Fisheries' *Sakumotsu Tōkei* [*Statistical Survey on Crops*], *Nōgyō Keiei Tōkei Chōsa* [*Statistical Survey on Farm Management and Economy*], and *Shokuryō Jukyūhyō* [*Food Supply and Demand Balance Sheet*].

Diminished sense of appreciation toward rice

Figure 2.1 shows the amount of rice harvested in 1 hour. As the yield per unit area grew and, concurrently, the time needed to harvest rice shortened, the amount of rice harvested per hour increased rapidly. In terms of the amount of rice a Japanese person consumes, in 1951, 1 hour's labor by a rice farmer could produce enough rice for five people to eat in a day. This jumped to 1 hour's labor producing enough rice to feed 12 people in a day in 1967, or 2.5 times the rate of 1951. The same figure had rocketed to 167 people in 2020.

The first turning point in 1967 was also around the time when the pace of increase in the amount of rice produced per hour began to gain momentum. And while the Japanese people's appreciation of rice farmers whose labor allows them to enjoy rice remained unchanged, it seems words of appreciation ceased to carry the same weight at around that time.

Ideal rice farming

Since 1955, and as rice farming yielded harvests sufficient to meet the need for rice with even some leeway in the supply, the ideal for rice farming was no longer confined to the prevention of starvation. On many fronts people became increasingly vocal about what ideal rice farming should aim at.

Take the Agricultural Basic Act, enacted in 1961, for instance. Nicknamed the 'constitution of the agricultural sector,' this act aimed at correcting the disparities in productivity, income, and living standard between agriculture and other industrial sectors that widened during the period of rapid economic growth. Under a policy called selective expansion, the act set a goal for aggressively expanding production of highly marketable agricultural and livestock produce such as high-end vegetables, while concurrently reducing production for silk farming and barley. As for rice, the act set a goal of reducing the cost of rice production. Even though the Japanese government had strongly encouraged rice farming during the food shortage, once it detected an easing in the food shortage, it laid out a new goal for rice farmers without waiting for the achievement of self-sufficiency in rice. As this incident shows, rice farmers in 1967 were in no mood to enjoy a sense of accomplishment over the attainment of rice self-sufficiency.

The process of converting the state of short supply of rice, even though all rice farmers had harvested rice in all of their rice paddies, to the state of rice sufficiency highlighted the disparities in efficiency among rice farmers more than ever before. Studying rice production in the Hokkaidō, Kinki, and Kyūshū regions in the year of the first turning point (1967), agricultural economist Kajii Isoshi found that the disparity in production capacity among rice farmers had widened to the extent that large-scale, highly productive rice farmers were able to expand their scale of operations by renting the rice paddies of small-scale farmers with low productivity. Based on this finding, Kajii pointed out that rice production in Japan had begun to change qualitatively in 1967. He pointed to the possibility that the disparity among rice farmers might promote the tendency for rice paddies to be concentrated in the hands of full-time farmers. It was around the same time that other agricultural economists, Sakamoto Kusuhiko and Imamura Naraomi, found that the disparity among rice farmers in the Kaetsu region of north-eastern Niigata prefecture merited special attention.

Some ideals for rice farming change their meaning when the supply of rice is sufficient. For instance, the ideal of making rice production more time efficient has become more about reducing the number of rice farmers, so as to increase laborers in other industries, or reducing the cost of rice production, than it has about making the time to produce more rice. Also, the gap between the ideal and reality, which was exposed when people could not wholeheartedly rejoice in the achievement of rice self-sufficiency, garnered more attention. Another gap was between the ideal of creating independent professional farmers through the permeation of agricultural machinery and rice paddy improvements, on the one hand, and the reality of the growing number of part-time farmers as rice production became easier, on the other. Also, more people than before became concerned about rice farming being heavily reliant on chemical fertilizers, which would exhaust the rice paddies.

Waning interest in rice began 100 years ago

A 2015 survey on the dietary lifestyle of Japanese conducted by the Ministry of Agriculture, Forestry and Fisheries revealed that as many as 20% of men in their twenties eat no rice in a month. While elders are liable to say, 'Today's young people don't eat rice anymore,' the situation is not entirely the younger generation's fault. Japanese started losing interest in eating rice about 100 years ago, which means that almost the entire population of Japan today has lived only in the age of disenchantment with rice.

Figure 2.2 illustrates changes in the Japanese population and the amount of rice the Japanese consume. The figure shows that, even though per-capita consumption of rice temporarily peaked in 1905, per-capita rice consumption began to decline as the general trend in the peak year of 1923, 3 years before the beginning of the Shōwa period (1926–1989). Interestingly, the figure for 1957, the postwar peak year, fell short of the prewar peak year. Although Japan entered the era of satisfactory rice supply that year and its citizens had lived through the hardship of the Pacific War, the Japanese people were still not eating as much rice as before. This indicates that Japanese had already started losing interest in eating rice toward the end of the Taishō period (1912–1926). It should be noted, however, that the amount of rice consumed by the Japanese as a whole continued to increase due to population growth. Subsequently in 1963, rice consumption by the Japanese as a whole peaked; it was from this point that the declining trend commenced. One reason the achievement of rice self-sufficiency was anticlimactic was because the per-capita decline in rice consumption had already started 44 years earlier, and it was followed by the decline in national rice consumption that had begun 4 years prior.

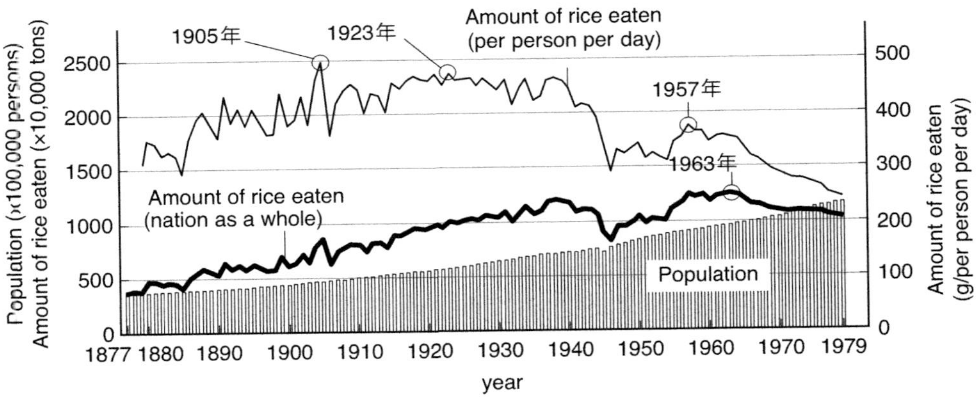

Fig. 2.2. Change in population and the amount of rice eaten. *Source*: based on the Bureau of Agricultural Affairs, Ministry of Agriculture and Forestry (1933); Research Division, Minister's Secretariat, Ministry of Agriculture, Forestry and Fisheries (1976); the Ministry of Agriculture, Forestry and Fisheries' *Shokuryō Jukyūhyō* [*Food Supply and Demand Balance Sheet*]; and the Bureau of Statistics, Ministry of Internal Affairs and Communications' *Jinkō Suikei* [*Population Estimates*].

Wheat consumption started in urban areas

Reasons why the Japanese became disinterested in rice differ from time to time. Particularly noteworthy about the timing of the Japanese people's disinterest in rice is the concurrent increase of wheat consumption. Consumption of bread and meat, which originated at the time of the Bunmei-kaika (meaning 'civilization and enlightenment,' or the phenomenon of Westernization in Japan) during the Meiji period, became increasingly popular, starting in urban centers. Comparing the consumption of staples per capita in 1955 between urban centers and rural areas, we see that only 92 kg of rice was eaten in urban centers compared with 158 kg in rural areas, while bread and noodles were consumed more in urban centers.

In the quarter century between 1935 and 1959, the proportion of rice as the staple declined from 86% to 76%, while that of wheat rose from 6% to 16%. In around 1967, now familiar brands of instant ramen noodles, including Sapporo Ichiban, Demae Icchō, and Myōjō Charumera, were marketed one after another, following the pioneering Chikin Rāmen. These foods led to the gradual permeation of wheat-based foods in the daily meals of Japanese.

Japan Continually Reliant on Food Imports

Japan's modern and contemporary development was reliant on imported cereals

Since the Meiji period, Japanese have not had the satisfaction of filling their stomachs with only domestically produced food. Japan's modern and contemporary society and economy have developed with imported food as the premise.

When poor harvests struck during the Edo period, each *han* (feudal lord-controlled domain) separately blocked food from flowing out, leading to uneven distribution of food throughout Japan as a whole. Because of this, people in those days were liable to suffer from famine. In the Meiji period, as the centralized state was founded and the system to distribute food to all corners of the country via railways was consolidated, the number of famine victims was reduced drastically.

Another reason why instances of death by starvation declined, we must remember, was that the Meiji government continued the open-door policy that the Tokugawa shogunate in the Edo period had instituted toward its end. The open-door policy allowed the emergency import of rice. In the second year of Meiji, when the price of rice rose rapidly due to the Boshin War (the 1868–1869 civil war between imperial and shogunate forces) and a poor harvest, the government immediately imported 100,000 tons of rice, and then another 320,000 tons. This second amount was equivalent to 9% of the rice produced in Japan in the ensuing year. Subsequently, as we saw earlier, Japan came to rely on rice from Taiwan and Korea. Then, after the Pacific War, Japanese became increasingly reliant on imported wheat.

Globally speaking, in around 1955, when Japan had not yet accomplished self-sufficiency in rice, agricultural produce all over the world went from being in short supply to being available in excess. The situation was favorable not only for Japan, which wanted to import agricultural produce, but also for countries with surpluses of agricultural produce. One such country was the United States, which had an excess supply of cereals.

Charitable activities of the United States and its food strategy

IN 1946, immediately after the Pacific War, when anti-Japanese sentiment was still rampant in the United States, Asano Shichinosuke, an American of Japanese descent, and his associates who were deeply concerned about the food shortage in Japan founded the Licensed Agencies for Relief in Asia (LARA) with the U.S. government's endorsement. A variety of charitable contributions by LARA, ranging from food such as wheat to clothes, medicines, and livestock, symbolized the dawning of a new era of United States–Japan friendship. The Miss Japan contest originated as a means to select a goodwill ambassador to convey the appreciation of the Japanese people to LARA. In subsequent years, as Japan and the United States continued to nurture their friendship through food aid, food aid began to gradually take on the connotation of a U.S. food strategy. In this way, Japan became entangled in the U.S. strategy.

The cereal supply situation of the United States

During World War II, the United States succeeded in drastically improving efficiency in wheat production to feed the soldiers of the Allies. When the July 1953 truce in the Korean War coincided with the timing of a good harvest, however, the problem of an excess supply of wheat all at once surfaced to the extent that even wartime cargo ships had to be used as storage. The U.S. government had to pay something close to ¥200 million daily for storage alone.

Troubled by surplus cereals, U.S. President Dwight D. Eisenhower introduced a new bill to Congress in January 1954 that proposed using surplus agricultural commodities – valued at U.S.$1 billion – to help the economies of friendly nations. Thirty-five market researchers were dispatched to various regions of the world to sell American agricultural produce. Japan was deemed the most promising market for wheat and other American agricultural produce.

Self-defense forces and school lunch borne out of U.S. food strategy

In June 1954, 5 months after Eisenhower introduced the agricultural commodities bill, the Self-Defense Forces Law was enacted in Japan. There was a close connection between this law and U.S. agricultural produce. Japan and the United States agreed that the United States would provide Japan with its surplus cereals for free, and Japan would finance the U.S. military bases in Japan and the founding of the Self-Defense Forces through sales of the cereals. This way, the United States could reduce military expenditures and dispose of surplus cereals, while Japan could upgrade its defense capability inexpensively.

Also, the new U.S. law allowed countries that could not import agricultural commodities due to a shortage of foreign currency to use their own currencies. Moreover, the law also included a provision allowing importing countries to hold the payment to use for their own economic development. Under the United States–Japan agreement and related legislation,

Japan imported 800,000 tons of wheat, 155,000 tons of barley, 100,000 tons of rice, 110,000 tons of animal feed, and non-fat milk, among other commodities. The total amount of agricultural produce entering Japan this way was worth ¥54.12 billion, when the national budget of the time was ¥1 trillion. One can appreciate how significant these imports were to Japan.

The School Lunches Act was promulgated and enforced at this time as well, on June 3, 1954. Originally, the Japanese government had aimed to legislate a school lunch law independent of the U.S. food strategy. The Ministry of Finance, however, was reluctant to agree with the Ministry of Education's plan to institutionalize school lunches using government funds. The U.S. strategy for its surplus agricultural commodities was truly a godsend; it played a decisive role in enabling the Ministries of Education and Finance to reach a compromise. To explain the thinking behind the proposal for the School Lunches Act, Minister of Education Ōdachi Shigeo said at the National Diet, the Japanese parliament, that it would be crucial for Japanese to familiarize themselves with food made from flour, such as bread and noodles, from childhood, to improve the rice-centric Japanese diet. School lunches centered around bread and non-fat milk began all over Japan, promoted by government-initiated propaganda that rice was inferior to bread. Japan and the United States agreed that Japan's school lunch system would use 185,000 tons of wheat per year for 4 consecutive years. Additionally, under the instruction of the U.S. government and business organizations, nutrition guidance vehicles traveled to all corners of Japan to promote the use of wheat by demonstrating how to cook home meals prepared with wheat and soybeans.

While the Ministry of Agriculture and Forestry had been reluctant at first to expand wheat imports or to popularize flour-centered meals among the Japanese, in the end it agreed to import wheat and decided to help develop the bakery business. Then Vice-Minister of Agriculture, Forestry and Fisheries Tōhata Shirō reminisced about the reasons behind this change of heart by the ministry. They included, aside from the acute shortage of food, the ministry's wish to finance the planned construction of the Aichi Canal and the Hachirōgata reclamation project through the importation of American agricultural produce without actually using foreign currency, which was tantamount to de facto introduction of foreign capital.

Japan's livestock industry developed on the assumption of animal feed imports

In January 1960, a U.S. military plane carrying 35 pigs landed in Japan. The pigs were sent as reconstruction assistance for the livestock industry of Yamanashi prefecture, which had been devasted by the Isewan Typhoon (Typhoon Vera) the previous year. This charitable act nicknamed Operation Airborne Pigs was a major project that was made possible through collaboration among the U.S. Department of Agriculture, the U.S. Air Force, and the National Corn Growers Association. Along with the pigs, 1500 tons of corn for animal feed was also delivered. Cultivating a new market channel for American corn in Japan was an ulterior motive for Operation Airborne Pigs.

The pigs brought to Japan were a large breed with high productivity, unlike the mid-sized breed prevalent in Japan in those days. Their arrival triggered a surge of momentum toward breeding highly productive pigs in Japan. By the time the last of the gifted pigs had died, they already had 500,000 offspring. Originally, Japanese farmers had usually fed pigs with leftover human food, but gradually the use of animal feed made with U.S.-produced corn replaced the old ways because the product was so easy to use.

For the Japanese, livestock originally were tools for cultivation and transportation. While it is not that Japanese had never eaten animal meat at all, animal husbandry as it is known today was not developed until 1872, when Emperor Meiji lifted the ban on meat eating. The ban had been imposed partly due to the influence of Buddhism and because of Emperor Tenmu's prohibition in 675 on the killing of animals and the eating of meat throughout Japan. In corollary, then, the animal feed industry also had not developed. It was on the assumption that animal feed could be imported in bulk that animal husbandry as an industry started to emerge and develop in Japan, supported by imported cereals. Thus, Japan became the world's greatest importer of American cereals.

'Constitution of agricultural sector' reflects U.S. intentions

The U.S. food strategy was built on the basis of the 'selective and efficient expansion' speech by the U.S. delegate at the 1953 meeting of the Food and Agriculture Organization (FAO). The United States requested other member countries to select and produce agricultural commodities other than those that the United States, problematically, had overproduced. This was one of the U.S. global strategies during the Cold War.

Japan's Agricultural Basic Act represented Japan's policy on selective expansion of agricultural products that was adopted in response to this U.S. call for selective and efficient expansion. Crops that could compete with U.S. agricultural produce were removed from the list of crops earmarked for expanded production. This is what led wheat production in Japan to a 'merciful death.' Even though the Agricultural Basic Act advocated the promotion of the animal husbandry industry, it did not explicitly name crops for animal feed that would compete with U.S. producers among those designated for expanded production.

Regarding rice, although the United States in those days produced some 2.5 million tons of rice, U.S. rice did not compete with Japanese rice because they were different breeds. Unlike the case of wheat and animal feed, therefore, rice was a commodity that the Japanese government could subsidize without displeasing the United States. As the dietary habits of the Japanese underwent changes, it would not have been unnatural if domestic production of wheat increased or subsidies for rice farming were reduced. Neither would it have been particularly surprising if the production of cereals for animal feed expanded along with the development of the animal husbandry industry. But history shows that these things did not happen. United States–Japan relations during the Cold War were a background influence.

Sudden drop in food self-sufficiency

In the 20th century, the United States succeeded in establishing a method to produce cereals in abundance for export. It coupled surplus wheat with military aid and/or sales of corn for animal feed with promotion of the animal husbandry industry in the importing countries. In contrast, Japan's self-sufficiency in cereals dropped abruptly from 82% in 1960 to 33% in 1980, a 49% point drop over 20 years.

Technically speaking, the rate of food self-sufficiency as it is often referred to in news media is the rate of comprehensive food self-sufficiency based on calorie intake. Because products from livestock fed with imported animal feed are not counted as domestic products, the more domestically produced meat, eggs, and milk that are consumed, the lower the food self-sufficiency rate becomes due to the animal husbandry industry's heavy reliance on imported animal feed. Japan's comprehensive food self-sufficiency rate based on calorie intake declined only 2 points in the 15 years between 2005 and 2020, from 40% to 38%. In contrast, the same rate declined as much as 20 points between 1965 and 1980, from 73% to 53%.

Control of Rice by the State

The price of rice

If the price of rice fluctuates through market mechanisms, so that it is high when rice is in short supply and low when rice is in excess, then both rice producers and rice consumers would become greatly concerned about self-sufficiency in rice. This has not happened in Japan, however, because the price of rice has been determined not by the market but by the national government.

To begin with, by nature the price of agricultural produce is highly liable to frequent fluctuation. Unlike industrial products, agricultural produce is greatly affected by nature; crops harvested are likely to increase or decrease more than planned. Because harvest seasons are fixed and most agricultural produce is highly perishable, the amount of produce marketed also fluctuates greatly. Nevertheless, consumers have no other choice but to purchase essential food no matter how expensive it is. Yet no matter how inexpensive produce becomes, one can only eat so much. Under these circumstances, the price of agricultural produce is likely to jump when supply is short and plummet when supply is abundant.

Of all agricultural produce, the price of such an essential food as rice is particularly liable to spike when the harvest is poor and fall when the harvest is good. This tendency is further aggravated by speculative buyers. Especially in a period like the Taishō era, when the tendency to rely on rice for daily meals was getting stronger while warehouse capacity was still inadequate, the range of fluctuation in the price of rice was liable to be notably wide.

Direct control of rice by the state

In the Meiji period, the 1873 Land Tax Reform required landowners to pay property tax in cash rather than goods. This expedited the rapid commercialization of rice. At this point, as far as the price of rice was concerned, government intervention was still quite relaxed. As the Taishō period began in the midst of wild fluctuations in the price of rice, however, the Japanese government suddenly switched to aggressively intervening in the price of rice.

In the early Taishō period when the rice price fell, the government attempted to regulate the amount of rice marketed and the price, by buying up rice and promoting construction of warehouses to store it. The government also established a research commission on rice price adjustment within the Ministry of Agriculture and Commerce. But the price of rice rose quickly, even before the commission could decide on a method. The cause was twofold. In addition to the big increase in rice consumption by the greatly expanded urban population caused by rapid industrialization, the imminent Siberian Intervention triggered speculative buyouts of rice. It was then Minister of Agriculture and Commerce Nakashōji Ren who responded to this situation. Formerly a prosecutor, Nakashōji attempted to solve the problem of speculative buyouts by arresting Oka Han'emon and Masuda Kan'ichi, two renowned big-time speculators. Still, the price of rice remained abnormally high.

The public mood in favor of the government regulating rice intensified with the surge in the price of rice that resulted in the 1918 rice riots. More than 1 million people are believed to have demonstrated, with some of them becoming violent. It is said that the riots spread to 369 locations in Hokkaidō, three major municipalities, and 37 prefectures. The social confusion in those days was so horrendous that the National Junior-High School Baseball Tournament (today known as the National High School Baseball Championship, or the Summer Kōshien) was abruptly cancelled after the players had already taken the field. In its history, the Summer Kōshien has been cancelled only three times. The other two times were the start of the Pacific War and the COVID-19 pandemic. This reveals the salience of the social confusion caused by the rice riots. In response, the government in 1921 enacted the Law on Rice to restrict the purchase and importation of rice when supply was overly abundant and promote the sale of government-owned rice and the importation of rice when supply was short.

Tighter regulation of rice by the state displayed its merit when in 1923 the government prevented possible social panic triggered by rice shortages by promptly transporting rice onboard a naval vessel from Osaka to Tokyo, which had been devastated by the Great Kantō Earthquake. Subsequently, after 1937, when the First Sino-Japanese War started, the government began to forcibly intervene with all commodities on a war footing. A system was established under which rice farmers were required to sell all the rice they harvested, excluding that for their own consumption and the seed crop for the next year, to the state. The government then controlled and rationed the rice thus collected. Other laws to strengthen the government's grip on rice continued to be enacted one after another, including one regulating the amount of rice to be transported from Taiwan and Korea. The rice exchange post system, Japan's system of rice brokers that had developed independently and privately in the Edo period, was also abolished around this time.

In February 1942, 2 months after the beginning of the war with the United States, the system for state control of rice was completed with enactment of the Staple Food Control Act under the slogan, 'Do not grieve scarcity, deplore inequity.' While, previously, the government had exercised indirect control over rice, this law allowed the state to directly control all of the rice in Japan. As a result, rice became shut off from market and international influences. Because white rice alone came under governmental control, other kinds of rice found in the rice section of supermarkets in Japan today, such as red rice, black rice, or haiga rice (rice with the germ intact), did not gain wide usage among Japanese consumers. Farmer Asahi Taneichi and his son Hajimu spent their entire lives – two generations of devotion – popularizing haiga rice. Hajimu reminisced that when only white rice was adopted for government rationing, 'Dad was in tears, mortified at the government's decision. But we can't resist the trends of time, can we?' (Terayama, 1985)

Attainment of affluence and the Staple Food Control Act

Under the Staple Food Control Act, the Japanese government decided everything concerning rice, from the selling price for farmers to the purchase price for consumers, the wholesale price, and the handling fee. Moreover, rice distribution was confined to a fixed route: beginning with the rice farmer, the rice went to the rice farmers' association, then the state, next to distributing organizations, and finally to the consumer. Noncompliance or shortcutting the route were prohibited. To buy rice, households had to present their rice and grain ration passbooks issued by the government. In this way, all rice distribution was rationed by the government institutionally until 1968.

As hinted by rice farmers' reluctance to deliver their rice during the food shortage immediately after World War II, the Staple Food Control Act was originally a policy to suppress the price of rice – to the rice farmers' disadvantage – so that Japanese consumers could eat inexpensive rice. The Staple Food Control Act regime also became a mechanism to support laborers during the postwar era of rapid economic growth with a secure supply of affordable food. Partly driven by the engine of inexpensive rice, Japan's gross national product caught up with that of France, Britain, and Germany one after another between 1966 and 1968, after the country achieved rice self-sufficiency. Japan eventually attained status as the world's second largest economy after the United States.

While the Staple Food Control Act brought food and economic affluence to Japan, rice farmers did not benefit. Even in a good harvest year, their income did not increase, and rice farming as an industry was not able to keep up with the growth of other industries. This was because the domestic rice price was set low by the Staple Food Control Act in order to offer rice to consumers at a low price. Other policies to offer rice cheaply also existed. For example, imported rice was so much more expensive than domestic rice in those days that the government spent more than ¥40 billion in subsidies to bring down the price of imported rice.

As of 1967, the Staple Food Control Act still retained the role of protecting consumers. The national government purchased rice at ¥650 for 5 kg. At this price, rice farmers with low-yield rice paddies would make no profit. The government purchase price was only a minimum price set to ensure that enough rice could be produced for the entire nation.

Reasons Behind Surplus Rice

Good harvest, poor harvest: always back-to-back

A rice shortage is a graver problem than a rice surplus. Therefore, for quite some time after rice self-sufficiency was achieved, the momentum to grow more rice remained unchanged. The backdrop to this were the vivid and bitter memories people had of being destitute at times of past poor harvests. Rice harvests being damaged by sudden summer cold snaps or for other reasons are common occurrences in Japan. The history of rice harvests per area shows an unstable fluctuating pattern (see Fig. 2.3). Besides, when the per-area harvest increases and per-capita consumption concurrently decreases, more of

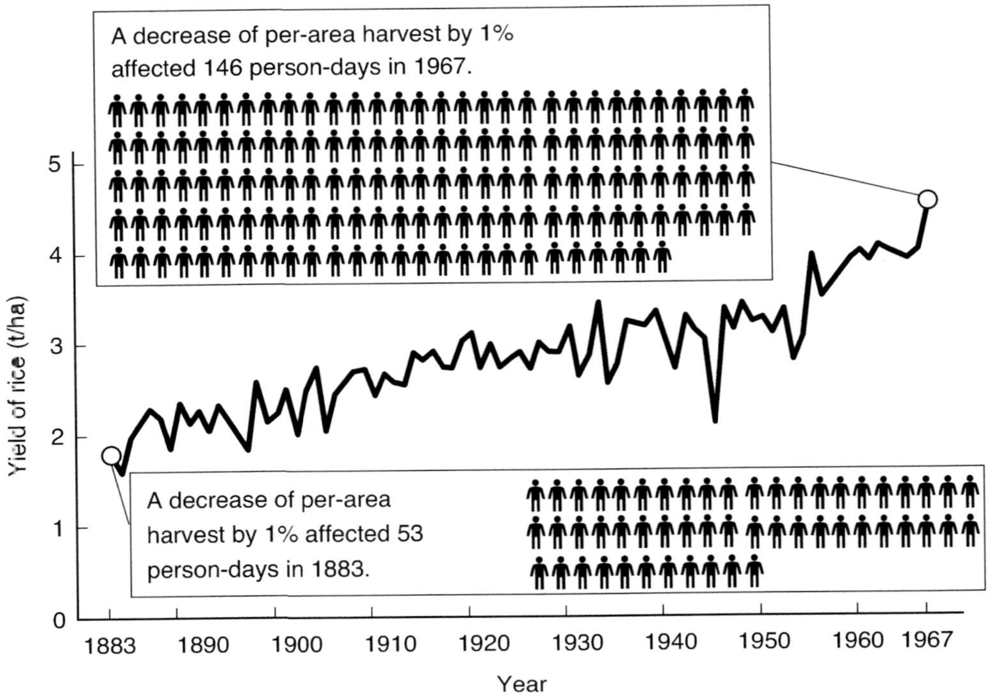

Fig. 2.3. Amount of rice crop per 1 ha and influence of 1% decline in crop yield. *Source*: based on the Ministry of Agriculture, Forestry and Fisheries' *Sakumotsu Tōkei* [*Statistical Survey on Crops*] and *Shokuryō Jukyūhyō* [*Food Supply and Demand Balance Sheet*].

the population is affected by good harvests or poor harvests. A decrease of per-area harvest by 1%, for instance, affected 53 person-days in 1883. In 1967, this jumped to 146 person-days, indicating a significant expansion in the influence from a poor harvest. This must have made policy responses much more difficult.

While the stockpile of rice immediately before the first turning point in 1967 increased from 52,000 tons in 1965 to 205,000 tons in 1966, this was still a mere fraction of the annual consumption of rice, which was 12 million tons. Incidentally, today the goal for the government rice stockpile is set at 1 million tons, an amount that could withstand 2 consecutive years of record-breaking, once-in-10-years poor harvests. Although rice self-sufficiency was indeed achieved in 1967, at the time 1 year's poor harvest would have directly caused a rice shortage. And, in fact, Japan imported 500,000 tons of rice in that very first year of rice self-sufficiency to prepare for just such an eventuality.

No business prospects for rice exports

Japan was once the third largest producer of rice in the world after the British Raj (India) and China, when Taiwan and Korea were counted as part of Japan. Nevertheless, Japan did not aspire to start exporting rice on a large scale. Authorities were, instead, of a more inward-looking mindset focused on somehow securing domestic sources of food for the nation. To paraphrase Fukuzawa Yukichi, an educator, philosopher, writer, and entrepreneur in the Meiji period, Japan was able to unabashedly boast of being 'Japan, the land of abundant rice' because Japan was isolated, having severed trade ties with many countries, and it did not know of the other numerous lands of abundant rice in the world. Globally, during Japan's early Meiji era, India was already exporting as much as 1.05 million tons of rice, while the United States was exporting 500,000 tons of wheat to Britain. These exports greatly affected the entire world market.

What has long remained unchanged is the high cost of rice production in Japan, making it difficult for Japan to export its rice. In pre-Pacific War days, Japan imported Rangoon rice (from Burma; now Myanmar) and Saigon rice (from French Indochina; now various countries in south-east Asia), both of which had been traded in London in 1933 at about 70% of the cost of Japanese rice. When the price of Saigon rice rose quickly in Marseille, France, in November 1936, Japanese rice was traded in Osaka at a price 1.6 times higher than Saigon rice. The amount of rice exported from Japan, where rice's cost price was high, was extremely limited. Given these situations, there was no vision on the part of Japan to pour energy into rice exports even when self-sufficiency was attained.

Was the rice surplus predictable?

While the amount of rice produced is steadily increasing, the amount of rice consumed has been declining. Moreover, business prospects for rice exports have been grim. Given these conditions, it was only a matter of time for rice to be in excess supply. However, both bureaucrats and academic experts in Japan failed to predict the timing of this inevitable occurrence in a credible manner. Although there already was a perception at the time of the 1961 Agricultural Basic Act advocating selective expansion that the future would see a rice surplus, it was beyond anybody's expectation that the day would come in less than 10 years.

In the days immediately before the first turning point of 1967, Japan had to import a total of 2 million tons of rice in 1965 and 1966. Added to this situation were such factors as much more rice consumed than expected by the postwar baby-boomer children who were in their 'hearty appetite' years; more than enough surplus rice in the hands of rice farmers who had kept the rice for family consumption; and heightened criticism in the Diet of the accuracy of the Agricultural Basic Law's forecast. All of these contributed to misjudgment of the timing for a rice surplus. This means that even though there was plenty of time between 1961, when the Basic Law was enacted, and 1967 to contemplate what should be done about rice after the imminent attainment of self-sufficiency, authorities continued to procrastinate on giving the question their earnest consideration. As a result, Japan had to deal with being confronted abruptly by a rice surplus problem.

Perceptions about an imminent rice surplus were not uniform among researchers, either. In a public debate on issues concerning Japan's agricultural industry convened in November 1965, immediately before the first turning point in 1967, positions varied. Some participants predicted that rice self-sufficiency was just around the corner. Others wondered how much rice Japan was likely to need in 10 years. Still others argued, 'Rice is not something that is produced so much because so much is needed. Instead, so much rice is used because so much can be produced' (Tōhoku Daigaku Nōgaku Kenkyūsho, 1966).

What is Rice to the Japanese?

Genesis: the treatises on rice and the Japaneseness boom

The achievement of self-sufficiency in rice unleashed a torrent of excellent works on relations between the Japanese and rice. Although some scholars in the past, including Yanagida Kunio and Oka Masao, called for caution around the assumption that rice is the one and only absolute existence for the Japanese, it was the general trend in prewar academia to perceive rice as such a thing. Moreover in the postwar days when the Marxist view predominated in the study of history in Japan, it remained fashionable for quite some time to understand Japanese history as centering around rice.

However, from the late 1960s to the 1970s, criticism suddenly erupted of the view of Japan's history being centered around rice. Some of the works that stand out include Miyamoto Tsuneichi's *Seigyō no Suii (Shifts in*

Bread-and-Butter Jobs; Miyamoto, 1965) and Nakao Sasuke's *Saibai Shokubutsu to Nōkō no Kigen* (*Domesticated Plants and the Origin of Agriculture*; Nakao, 1966) in the 1960s, as well as Sasaki Kōmei's *Inasaku Izen* (*Before Rice Cultivating*; Sasaki, 1971) and Tsuboi Hirofumi's *Imo to Nihonjin* (*Potato and the Japanese*; Tsuboi, 1979) in the 1970s. Archaeologist Andō Hiromichi points out that the critical juncture in this shift in intellectual climate was around 1968. In 1985, Ueyama Shunpei and Watabe Tadayo co-edited *Inasaku Bunka* (*Rice Cultivation Culture*; Ueyama and Watabe, 1985), in which they claimed theirs was the first book to refer to rice cultivation culture and pointed out that a mere 20 years or so had passed since this term was accepted by academia.

Although the Japanese have had a long history of association with rice, it wasn't until after the attainment of rice self-sufficiency that people became clearly conscious of the relationship. This heightened awareness emerged because people realized, at the time rice self-sufficiency was achieved and then as rice surpluses appeared, that rice is by no means the one and only absolute existence. Had the Japanese refused to eat wheat and continued to rely solely on rice, rice might have retained its special position. That would have given birth to a whole different school of treatises on rice and Japaneseness. When rice, which seems to be deeply related to the Japanese identity, became superfluous, it provided the Japanese people with an opportunity to ask themselves what rice is to them and, ultimately, to take another good look at their self-perception.

Rice self-sufficiency's influences on natural scientists

Continued rice surpluses not only affected scholars in the humanity and social sciences, they also confronted natural scientists with new questions. In 1967, there were many scholars in Japan who had entered the world of academia in the midst of the severe food shortages during and after the Pacific War. To these scholars, it was 'utterly unthinkable' when they started studying agriculture professionally 'that the day will come in Japan when we see a rice surplus' (Sugawara, 1980). Tsunoda Shigesaburō, an expert in rice breeding, introduced himself once at an academic meeting with these words: 'I specialize in plant thremmatology—high-yielding breeding to be more precise. This might make me one of the villains behind today's rice surplus' (Tsunoda, 1981). Attempting to harvest as much rice as possible was once an absolutely 'correct' behavior both from the viewpoint of scientific advancement and as an effort to make society a better place. When people realized at the time rice self-sufficiency was attained that scientific advancement and social improvement are not always accomplished simultaneously, however, what had previously been believed in unconditionally turned into something to question.

The Year 1967 in Retrospect

Japan achieved self-sufficiency in rice in 1967. For the Japanese people who had long cultivated rice, it was a historic accomplishment that reflected the ardent wishes of generations of Japanese. It was the attainment of a truly national goal. This accomplishment marked not only an important turning point in postwar rice policy but also a historical turning point in rice farming since its genesis in the Jōmon period.

In retrospect, 1967 marked a watershed for rice farming as society shifted from being on the verge of starvation to being liberated from the fear of starvation. After that year, rice farmers began to pursue all kinds of ideals other than the prevention of starvation. Yet we cannot say that the Japanese people in those days were aware of this shift. Views on a future rice surplus were not uniform even among bureaucrats or scholars. As a result, rice farming in Japan lost its bearing for several years, being deprived as it was of any vision beyond accomplishing self-sufficiency in rice. As the years passed, and as the rice stockpile continued to expand, the presentiment regarding a rice surplus gradually turned into conviction. Ten years later, in 1978, Japan's rice farming experienced its second turning point.

References

Food Agency (1980) *Nippon no Shokuryō Shigen—Nipponjin to Kome no Rekishi* [*Japan's Food Resources—History of the Japanese People and Rice*]. Chōū Tsūshinsha, Tokyo.

Miyamoto, T. (1965) *Seigyō no Suii* [*Shifts in Bread-and-ButterJobs*]. Kawade Shobō Shinsha, Tokyo.

Nakao, S. (1966) *Saibai Shokubutsu to Nōkō no Kigen* [*Domesticated Plants and the Origin of Agriculture*]. Iwanami Shoten, Tokyo.

Nihon Keizai Shimbun (1967) Kisha techō [Reporter's memo]. August 30.

Sasaki, K. (1971) *Inasaku Izen* [*Before Rice Cultivating*]. NHK Books, Tokyo.

Sugawara, T. (1980) Shiiku mai no seisan gijutsuteki kanōsei ni tsuite [On production technological possibility of rice for animal feed]. In: Nōrin Suisan, G.J.K. (ed.) *Kome No Shiryōka Wo Kangaeru* [*Discussions on Animal Feed Made of Rice*]. Nōrin Suisan Gijutsu Jōhō Kyōkai, Tokyo.

Takase, K. (1977) *Kome no Hishi* [*Hidden History of Rice*]. Marunouchi Shuppan, Tokyo.

Tanaka, S. (1974) *Nippon Sensō Keizai Hishi* [*Hidden History of Japan's War Economy*]. Hara Akira (ed.), Tanaka Shin'ichi Nippon Sensō Keizai Hishi Kankōkai, Tokyo.

Terayama, Y. (1985) *Zoku Anotoki Konohito* [*That Time This Person: Part 2*]. Rakuyū Shobō, Tokyo.

Tōhoku Daigaku Nōgaku Kenkyūsho (ed.) (1966) *Inasaku Gijutsu Hatten no Ronri to Hōkō* [*Logics in and Direction of Rice Farming Technological Development*]. Nōsan Gyoson Bunka Kyōkai, Tokyo.

Tsuboi, H. (1979) *Imo to Nihonjin* [*Potato and the Japanese*]. Miraisha, Tokyo.

Tsunoda, S. (1981) Kome no seisan chōsei ni taisuru gimon to hihan [Questions and criticisms on rice production adjustment]. *Nōgyōhō Kenkyū* 15 and 16.

Ueyama, S. and Watabe, T. (eds) (1985) *Inasaku Bunka* [*Rice Cultivation Culture*]. Chūō Kōronsha, Tokyo.

Wada Hiroo Ikō shū Kankōkai, A. (ed.) (1981) *Wada Hiroo Ikō shū* [*The Writings Left by Late Wada Hiroo*]. Nōrin Tōkei Kyōkai, Tokyo.

3 From Rice Surplus to Paddy Surplus: The Second Turning Point in 1978

Abstract

Chapter 3 defines the year 1978 as the second turning point, when the vision following rice self-sufficiency was explicitly spelled out for the first time. It was envisioned that rice production should be curbed and production of other, non-rice produce should be encouraged in rice paddies. This vision highlights a uniquely Japanese feature of rice policy. The chapter reviews how the policy to export excess rice was abandoned under pressure from the United States and the circumstances around the decision to stabilize rice demand and supply through production adjustment. Chapter 3 also depicts how the use of non-rice-producing rice paddies failed to be provided with a sense of direction due to the absence of a full-scale legal framework, in spite of recognition of the importance of protecting rice paddies and against the backdrop of the rise in international concern about food security.

Rice Surplus Problem

Rice farming supported by tax

Since ancient days when political leaders and governments levied rice taxes on villages, rice has long supported Japanese society as a pillar of taxation. At the same time, those in power have continued to promote rice farming as a form of public works. While the balance between the two may vary from time to time, both rice and society have supported one another. When rice became superfluous, it became clear that all of Japanese society had supported rice farming through the tax system. When the policy frameworks restricting rice production, including the area reduction and production adjustment programs that linger on even today, were completed in 1978, the essence of the rice problem in Japan shifted from a surplus of rice to a surplus of rice paddies. Thus 1978 marks the second turning point for rice. Since then, rice farming in Japan has been supported by taxes.

First, let's review the process through which the rice surplus expanded the Japanese government's deficit. The price at which the national government buys rice from rice farmers is stipulated by the 1942 Staple Food Control Act. Over time, the act began to take on the color of a mechanism to narrow the widening income gap between urban laborers and rural farmers seen during the high economic growth that has occurred since the 1960s. After rice self-sufficiency was attained at the first turning point in 1967, the act was decisively transformed from a mechanism to secure rice during rice shortages to a scheme to protect the income of rice farmers. Raises in the national government's rice price thus became a means for the government to control the income of rice farmers.

Due to political considerations, the government took time to raise the price at which consumers purchased rice. This caused the government to begin to run a deficit by trading rice. This deficit was not problematic as long as a proper balance could be maintained between the burden on taxpayers and the merit of inexpensive rice that consumers enjoyed. Seeing as rice self-sufficiency had not been accomplished between 1960 and 1967, and Japan itself was in the midst of high economic growth, this government deficit could be tolerated for the time being.

The story changed, however, when rice self-sufficiency was achieved and, moreover, the massive rice surplus persisted. On top of shouldering increasing expenses for collecting and storing rice, the government also had to purchase rice even with no prospective buyers. The deficit swelled in no time. In 1970, the rice surplus reached an unprecedented 7.2 million tons, a colossal figure superseding the amount of rice rationed annually in those days. In 1971,

DOI: 10.1079/9781800627093.0003

the deficit caused by rice reached 55% of the annual budget of the Ministry of Agriculture, Forestry and Fisheries, far beyond a tolerable level. *Kome* (rice) was grouped together with two other causes of huge deficits, *kokutetsu* (the Japan national railway companies) and *kenkō hoken* (the national health insurance system). Referred to as the '3K problem' or the '3K deficit,' these three often became targets of criticism at the National Diet.

Cost of disposal of surplus rice

To dispose of the gigantic amount of surplus rice, some people even suggested dumping rice into the sea. In the end, though, the government decided on a method that would save incurring a cost for trashing the rice. Surplus rice was disposed of as an export item, as animal feed, and as an ingredient for processing into other food products. In 1971, the government bought rice at ¥710 per 5 kg and sold it at ¥615, incurring a deficit of ¥95 per 5 kg. In contrast, prices of rice for other uses per 5 kg were much cheaper, including ¥350–400 for processing rice, ¥200–250 for export rice, and ¥100–150 for animal feed. Thus, the government incurred a deficit of some ¥300 in the case of rice for processing and as much as ¥600 in the case of rice for animal feed. These were by no means negligible figures for a deficit the government incurred only through the buying and selling of rice.

Agricultural economist Saeki Sanae calculated the government's deficit per 5 kg of surplus rice at, tentatively, ¥675–1500, including transaction and administrative costs. This means that, with every 1 ha of rice paddy producing surplus rice, the government's deficit would increase by ¥620,000–1,380,000. This situation continued between 1971 and 1984, during which time the government deficit swelled to ¥3 trillion.

Unprecedented rice exports

About 40–50 years ago, Japan exported the largest amount of domestically produced rice in history. In those days, surplus rice was stored for a few years before being exported. Stores in 1971 amounted to 800,000 tons; in 1979 stores reached 900,000 tons. Japanese rice was exported mainly to Korea, Indonesia, and Bangladesh as part of economic aid in the form of loans with deferred payment allowances. It was a kill-two-birds-with-one-stone measure for Japan, which wished to deal with the surplus rice problem as well as contribute to solving the overseas famine problem.

The sudden rice exports from Japan, which had not been a leading rice exporter before then, at a lower, government-subsidized price, were tantamount to dumping. It meant that Japan's exports, though claimed as economic aid, were in fact a form of unjustified bargaining that disregarded international rules. The situation was an extremely annoying development for traditional rice exporters. When it heard of the American rice exporters' campaign to take the issue to the Office of the United States Trade Representative in 1979, the Japanese government decided to settle the issue by promising to restrict rice exports. Consequently, the export of rice as a form of overseas economic aid failed to become a pillar of the solution to the rice surplus problem.

Black market rice spotlighted

Although rationing is less efficient than free trading at market, it is an effective method when a commodity is in short supply. Thus, while rice was in short supply, introduction of the market mechanism was called off every time it was envisaged. After rice self-sufficiency was achieved in 1967, however, rationing's inefficiency became increasingly conspicuous, leading to studies on how best the two methods of rationing and leveraging the market mechanism could coexist. Exploration on the matter got into full swing after Minister of Agriculture and Forestry Nishimura Naomi declared at an October 1968 meeting of the Research Commission on Comprehensive Agriculture Administration of the ruling Liberal Democratic Party that 'free rice' would be officially endorsed under certain conditions. Free rice was the official name for black market rice.

Finally in 1969, rules allowing rice traders to shortcut government trading routes, the use of which had been mandatory until then, became official. While the effect of allowing black market rice into the mainstream market was still unpredictable, the government nonetheless promoted permeation of black market rice. The government even proposed subsidies for this purpose, because more black market rice in circulation meant a reduction in the amount of rice the government had to handle and, thus, a smaller deficit. Sugiyama Katsumi, the then director of planning of the Food Agency, and his deputy Yoshikuni Takashi reminisced in later days that the ulterior motive of the government was to gradually introduce the market mechanism to the rice trade by endorsing black market rice under certain governmental controls.

Around the same time, the government began to restrict the amount of rice it purchased. And the rice not bought by the government above the confinement was permitted to be distributed through the same route as the officially endorsed black market rice. Moreover, the government ceased regulating the price at which consumers purchased rice, a practice that had been in effect since the tumultuous days immediately following the Pacific War. With these acts, the stringent rice rationing scheme became a thing of the past, and rice, which had once been military goods, increasingly took on the air of a commercial commodity.

Black market rice carried in broad daylight on Kyoto Station platform (*The Mainichi Shimbun*, March 24, 1960). (Used with permission from Mainichi Shimbun)

Record of Diet proceedings altered in return for five or six bottles of sake

Behind the apparently contradictory coexistence of the Staple Food Control Act, by which the state strictly controlled rice, and the officially endorsed black market rice, was a uniquely Japanese legal consciousness. As legal scholar Kawashima Takeyoshi pointed out in *Nihonjin no Hō-Ishiki* (*Legal Consciousness of the Japanese People*) published in 1967 (Kawashima, 1967), law can be compromised little by little when a law and the reality are in contradiction. The legal consciousness of the Japanese does not criticize this attitude. On the contrary, it is inclined to be praised as a flexible handling of the contradiction.

Large numbers of consumers reached for black market rice both during and after the Pacific War. Illegal siphoning of rice became rampant, and some people realistically perceived black market rice as a necessary evil. By that time, the officially rationed rice had already been widely recognized as being unpalatable and many rice dealers purchased black market rice directly from rice farmers through brokers. Dealers also sold only the tastiest portions of rationed rice as black market rice at a high price. The dual realities of a huge surplus of rice on top of these illegal activities impacted the law enough to exact a compromise. Higaki Tokutarō, the then director-general of the Food Agency, reminisced that the allegory of a bag of surplus rice set on the equator going around the globe one and a half times was highly effective in convincing officials of the need to formally endorse black market rice.

Of course, there were critics of black market rice. One of them was Takatsuji Masami, the then director-general of the Cabinet Legislation Bureau and 'watchdog of law,' who pointed out at the Diet's budget committee that official endorsement of black market rice was problematic. Food Agency Director-General Higaki refuted that by saying, 'Mr Takatsuji is director-general, but so am I. I am in charge of administering three laws: the Staple Food Control Act, the Staple Food Control Special Account Act, and the Agricultural Produce Inspection Law. When the director-general in charge of these laws declares that official endorsement of black market rice is not a violation of the Staple Food Control Act, I expect to be trusted.' Hearing Takatsuji's rebuttal, Prime Minister Satō Eisaku laughed out loud. Later on, Satō said to Higaki, 'I have been prime minister for quite some time, but you are the first one to put on an impressive show of declaring that the director-general of the Cabinet Legislation Bureau and the director-general of the Food Agency are on equal footing' (Higaki and Terayama, 1998).

Higaki's rebuttal, however, cannot be found in the record of Diet proceedings. According to Higaki, a few officials of the Food Agency's planning division visited the Proceedings Department of the House of Councilors with five or six bottles of sake and altered the record. In any event, official endorsement of black market rice, which must have been in contradiction with the Staple Food Control Act, was realized without amending the stringent control act.

Solutions to the surplus rice problem

Disposing of surplus rice or reducing the amount of rice the government handles are only therapies addressing symptoms of the rice surplus problem. When exporting rice is not a feasible strategy with strong results, that leaves two means of fundamentally preventing rice surpluses. One is to increase the amount of rice consumption. The other is to control the amount of rice production so as to avoid oversupply.

School lunch programs started serving rice in the 1970s. But that outlet for rice was not limitless; the amount of rice consumed in school lunches began to dwindle in 1973. During that period, the need for population control was stressed, including at the 1974 First Japan Population Conference supported by the Ministries of Health and Welfare and Foreign Affairs that appealed for limiting the number of children per family to two. With the last of the baby boom nearing an end, the baby bust instead commenced from 1975.

In the late 1960s, international capital movement became liberalized in several steps. Starting in 1970, major American fast-food chains and family restaurants entered Japan one after another. Around the time of the 1964 Tokyo Olympic Games, private ownership of automobiles increased rapidly. This, together with the concurrent construction of expressways, facilitated the swift spread of domestic and overseas chain restaurants, particularly along main roads in the suburbs.

Due to a declining birth rate and diversification in people's eating habits, rice, which had been eaten in abundance before, became less popular. As the years passed, the premonition that the amount of rice eaten might not rebound turned to a conviction. From this point on, measures to cope with the rice surplus shifted in full swing toward controlling the amount of rice to be produced so as to avoid oversupply.

Start of the Policy to Curb Rice Production

Beginning of rice paddy surplus

Three months before his announcement on the official endorsement of black market rice, Minister of Agriculture and Forestry Nishimura Naomi issued a directive on the promotion of a comprehensive agricultural policy and instructed that study begin on a new policy. Accordingly, a goal was set to forgo rice farming in 10,000 ha of rice paddy during 1969, with each prefectural government allotted a mandatory portion of the total area. This was the beginning of the so-called area reduction and production adjustment policies. Cooperating rice farmers received subsidies commensurate to the paddy area left uncultivated – in other words, surplus rice paddies – reflecting the strong air of coercion accompanying the policy. In 1970, cooperating farmers received ¥35,000 per 10 ares, an amount that rivaled the initial salary of ¥39,900 per month for a college graduate office worker. As mentioned, black market rice was officially endorsed in 1969. Higaki Tokutarō, who, as director-general of the Food Agency, had endorsed black market rice, was appointed as vice minister of agriculture and forestry that same year. In later days, he confessed, 'Coercive allotment of acreage under the acreage reduction policy was really an emergency measure. While I was aware that it was a historic maladministration, there was no other way out when the issue of surplus rice had already emerged as being imminent' (*Mainichi Shimbun*, 1994).

A subsidy provided to a rice farmer who complied with area reduction would increase the financial burden on the government by ¥35,000 per 10 ares, or ¥350,000 per hectare. Still, providing the subsidy would be less costly than disposing of surplus rice (¥620,000–1,380,000 per hectare, as already mentioned). Nevertheless, it was only natural that some questioned the wisdom of this measure, saying, 'If a subsidy is to be provided anyway, then subsidized farmers might as well plant something else in dormant paddies.' Subsequently, a new scheme gradually formed in which the subsidy was provided on the basic condition that surplus paddies be planted with non-rice crops.

Rice farming's sudden downturn from the apex

The policy to curb rice farming commenced in 1969. That year was also the apex of rice production in Japan, seeing the highest records in the country's history: nationwide there were 3.44 million ha of rice paddies, of which 3.17 million ha were actually planted with rice. These figures are 1.4 times and 2.1 times, respectively, today's areas. Rice paddy area increased 120,000 ha in the 3 years between 1967 and 1969. Because this figure included the area of rice paddies displaced by cultivated dry fields, it does not necessarily represent the net increase in rice paddy area. In any event, the area increase was larger than the rice paddy area actually planted with rice in 2021 in Niigata prefecture (approximately 100,000 ha) and almost reached that for the entire island of Kyūshū (150,000 ha).

Although the government had once encouraged cultivation of rice paddies, it changed its policy in 1969. Instead, the government frequently issued circulars discouraging both the cultivation of new rice paddies and the conversion of cultivated dry fields to rice paddies. Countless development projects floundered in this era of surplus rice paddies that forced the alteration or abolishment of land use plans. Examples include the reclamation of Lake Hachirō in Akita prefecture, the reclamation of Fukushimagata lagoon in Niigata prefecture, the reclamation of Kahokugata lagoon in Ishikawa prefecture, the land improvement project in Nakaumi lake in Shimane and Tottori prefectures, and the reclamation of Isahaya bay in Nagasaki prefecture. The compensation issues and influences on the surrounding environments in these locations have not been settled to this today.

Some 150 farmers directly confronted the agricultural ministry's director-general of structural reform with a plea to grow rice in the reclaimed Fukushimagata lagoon in Niigata prefecture (*The Asahi Shimbun*, June 23, 1976). (Used with permission from Asahi Shimbun)

World food crisis in the 1970s and Japan's rice

In a speech given during the first year of the 21st century, U.S. President George W. Bush said, 'Can you imagine a country that was unable to grow enough food to feed its people? It would be a nation that would be subject to international pressure. It would be a nation at risk' (Bush, July 27, 2001).

Today, international common sense holds that agriculture is deeply connected with a country's security. This perception evolved in the 1970s, which saw a worldwide poor harvest in 1972 and an increase in crop prices caused by a huge purchase by the Soviet Union. Poor catches of anchovy for animal feed that year also aggravated the situation. With the worldwide spread of trade in agricultural commodities occurring at a time when Western countries were careful to avoid overproduction, the entire world experienced the food crisis simultaneously. In 1973, the United States suddenly banned the export of soybeans, causing steep rises in tofu (bean curd) and *natto* (fermented soybean) prices in Japan.

Having learned the lesson from this experience – namely, that when food was in short supply, each country without fail gave first priority to its

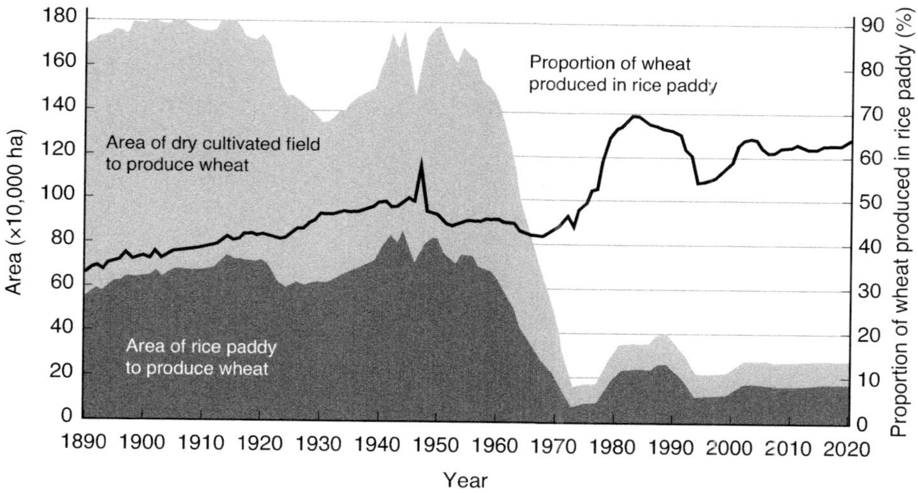

Fig. 3.1. Changes of area where wheat is produced. *Source*: based on the Ministry of Agriculture, Forestry and Fisheries (various years).

own people – countries worldwide came to recognize the importance of self-sufficiency in food. The U.S. embargo on soybeans in January 1980 as an economic sanction against Russia for its invasion of Afghanistan was most devastating. Now food was no longer a mere commodity. It had become the third weapon in international relations next to military arms and oil. In Japan, the term *self-sufficiency capability* was introduced for the first time in 1975 by a Ministry for Agriculture and Forestry document. This was followed in 1980 by a Diet resolution on the strengthening of food self-sufficiency capability. Subsequently, the perception that the surplus rice paddy subsidy would contribute not only to mitigating the rice surplus problem but also to enhancing food security took root in Japan.

On the other hand, wheat and soybeans, which had been sentenced to a merciful death, so to speak, suddenly entered the spotlight in Japan (see Figs 3.1 and 3.2). Production of wheat, which is a winter crop, rapidly decreased in the 1960s before picking up slightly through the expansion of its proportion of crops grown in rice paddies. Because soybeans are a summer crop like rice, and also are vulnerable to soil with a high water content, soybeans had hardly been grown in rice paddies previously. Today, however, more than 60% of Japan's soybean crop is being grown in rice paddies.

Breakthrough brought about by measure to reorganize rice paddy use

In response to the problem of surplus rice paddy area and the world food crisis, on April 6, 1978 the Japanese government issued a new policy named the Measure to Reorganize Rice Paddy Use. Prefectural governors nationwide received notification that same day. This was the second turning point in relations between rice and the Japanese. The adoption of this policy signified widespread recognition that restriction of rice production would become necessary in the mid to long term.

Past Japanese policies to discourage rice production were mostly emergency measures with a time horizon of one to, at the most, a few years. Farmers were encouraged to grow other non-rice crops in rice paddies based on the idea that the paddies were only temporarily dormant. Under the new policy, the central idea shifted to reorganization of rice paddy use and even agricultural methods themselves over a time span of 10 years or so on the assumption that the surplus of rice paddy area would continue for the mid to long term.

This new policy gave the government a framework for making genuine efforts to restrict rice production. This framework, however, had

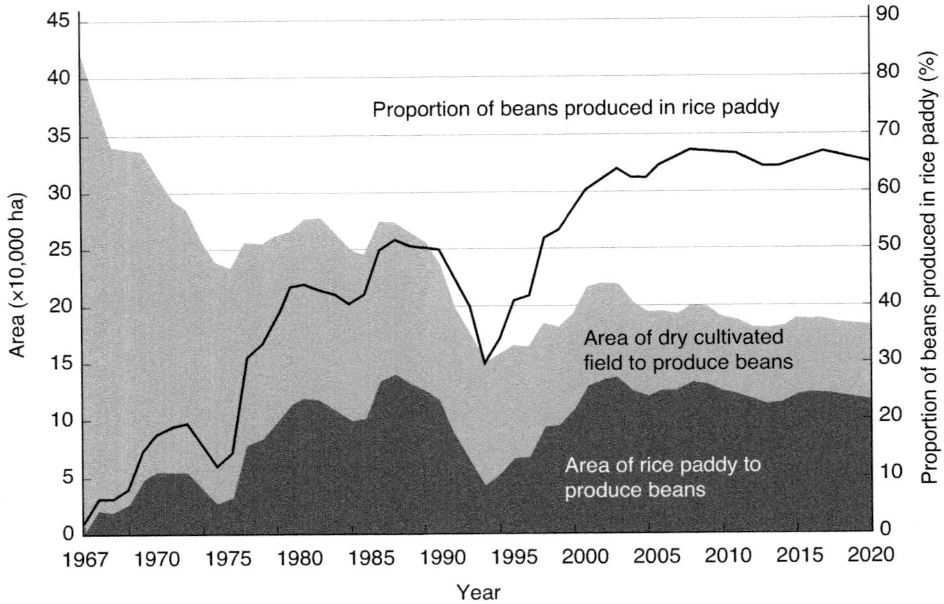

Fig. 3.2. Changes of area where beans are produced. *Source*: based on the Ministry of Agriculture, Forestry and Fisheries (various years).

a grave defect. This is because the policy was decided by a cabinet meeting. A highly important policy though it was, it did not go through a legislative process and, therefore, it had no legal basis. It was treated as a policy whose decision-making could be handled within the jurisdiction of the minister of agriculture and forestry. Because this policy to curb rice production was not supported by a law, its details were subject to frequent modifications. As such, it came to be passed on to future authorities as a highly complex policy.

A mid- to long-term policy also responsive to emergency

Ōga Keiji, who entered the Ministry for Agriculture and Forestry in 1967, the first turning point year for rice, as we have seen, reminisced in later days that, as a young official of the Policy Planning Office of the Minister's Secretariat, he had always felt dubious about the policy's defect. Had the policy merely been a short-term emergency measure, its legal position to restrict rice production would have been understandable. It was uncanny, however, for a policy with a mid- to long-term time horizon to play a major role without a solid legal basis of its own. Moreover, because the policy monitored the balance between the amounts of rice produced and rice consumed annually instead of every few years, it retained the unique capability to instate emergency measures to restrict rice production when the balance tipped toward oversupply even though officially the policy aimed at a 10 year vision.

Although this policy was finally provided with a respectable legal basis in 1995, as we will see in Chapter 4, the obscure nature of its legal framework still remained. The inherent contradiction in the policy's preparedness to annually contemplate measures to deal with rice paddy area surpluses while predicting the mid- to long-term persistence of the problem remains unsettled to this day.

Area reduction or production adjustment is not coerced

Many readers may have a vague perception that policies to curb rice production – that is, area reduction and production adjustment measures – were coercively imposed on rice farmers. In fact, some textbooks and media reports have explained these policies as such. In order to understand subsequent history without confusion, however, it is crucial to know that these policies were not coercive. The misperception exists because the government induced rice production reduction so forcibly. Coercion, however, is a world of 100% or zero with no in-between; the government's area reduction and production adjustment measures merely had an air of coercion.

This may not be a felicitous analogy, but the government's appeals during the peak of the COVID-19 pandemic for people to stay home, get vaccinated, and wear masks should still be fresh in readers' minds. The decision to comply or not with those appeals was left to individual citizens. The government did not coerce people but merely asked for compliance.

It takes a carrot-and-stick approach to induce people to comply with appeals of this kind. Carrots during the pandemic were, for example, economic impact payments to mitigate the pain of self-imposed sheltering at home or discount campaigns aimed at those who showed a vaccination certificate. The carrot for compliers de facto becomes a stick for noncompliers, while the stick for noncompliers de facto is a carrot for compliers. The carrot and stick are two sides of the same coin, and the larger or more intense the carrot or stick becomes, the more coercive air an appeal takes on until it de facto becomes a coercion. Policies to curb rice production were full of these carrots and sticks.

Nationwide, approaches to vaccine administration differed from region to region. On the individual level, some people found it uncomfortable to be maskless when everyone else was masked, while others may have been irritated by maskless individuals. Because the policy to curb rice production was not coercive either, its administration varied from region to region. Moreover, the framework of the policy was such that noncompliance could bring troubles to neighboring rice farmers in the same community. These carrot-and-stick arrangements generated numerous local responses, including administrative guidance on site and peer pressure within a community. On top of the appeals and subsidies by the national government, each municipal and city/town government as well as local agricultural cooperatives also offered their own guidance and subsidies, contributing to the increasingly stronger air of coercion on site.

Some people may misperceive the economic impact payments delivered during the shelter-in-place period of the pandemic to be compensation for compliance with the government's appeals. Equally, subsidies paid to owners of surplus rice paddy area tended at first to be regarded as compensation for compliance with coercion. This misperception regarding compliance came about because the subsidy offered under the policy to curb rice production was so high. Subsequently, however, the subsidy was proven to be too low to allow farmers to grow non-rice crops in all of their surplus rice paddy area.

Had the state forced farmers to cut back rice production, a tremendous amount of compensation would have been necessary. In reality, even though the rice reduction policy was not a coercive measure, the subsidy for surplus rice paddy area was so insufficient that rice farmers themselves had to contribute to mutual aid funds in order to mitigate the sense of inequity among rice farmers over the allocation of surplus rice paddy area.

Surplus rice paddy as a stick

Although farmers had been repeatedly notified of the measure to restrict cultivation of new rice paddies since 1969, new rice paddy area had actually increased. To remedy this situation, the government was thorough in its pursuit of a rice paddy usage reorganization measure as a stick. However, the Ministry of Agriculture and Forestry itself was aware that it would not be appropriate to limit cultivation of new rice paddies with a law because that could be tantamount to a restriction of private rights and provoke a constitutional crisis. The ministry explicitly called on local governments to pay attention to this risk. At the same time, the national government instructed local governments to thoroughly carry out such measures as making rice farmers who cultivated new rice paddies ineligible for public loans and doubling the rice paddy area in which farmers were requested to abstain from planting rice.

Aside from adjusting the rice paddy area to be left unplanted, the national government also ruled that any surplus area goals municipal governments failed to meet in a year would be added to the goals for the ensuing year. As for the allotment of surplus rice paddy area among municipalities, the national government explained that such was by no means a pressure, but merely a rule to ensure equality between those municipalities which cleared the goal and those that did not. Nevertheless, while the policy to curb rice production brought carrots in the form of government subsidies to complying municipalities, in actuality it could also impose sticks in the form of additional allotments of surplus rice paddy area to those prefectures, cities, and towns which overproduced rice.

Thus, using surplus rice paddies as a stick, the policy to curb rice production continued to be carried out, with the policy deliberately nurturing a mood of mutual monitoring and restriction among rice farmers. Some representatives of local farmers at the end of the chain of command were made to play the role of pleading with fellow rice farmers to refrain from growing rice. And some of them felt such pressure over being caught between their duty and their non-compliant rice farming neighbors that they killed themselves.

The duck's underwater paddling

The national government was once sued for compensatory damages caused by its policy to curb rice production on the grounds that the policy violated the people's right to grow rice. The suit was dismissed on the grounds that the policy did not prohibit or restrict growing rice itself. On the other hand, Mr F., a rice farmer in Toyama prefecture, was sued for going against the policy to curb rice production by marketing his rice harvest on the black market, which constituted a violation of the Staple Food Control Act, though Mr A. was not condemned for ignoring the policy itself.

Although the policy to curb rice production was not coercion by the state, it was in actuality a policy with a strong air of coercion. It may appear on the surface that the policy was launched smoothly, but underneath the surface there were numerous movements and actions all over Japan that are not recorded in research papers or documents. According to agricultural economist Arahata Katsumi, a prefectural governor in eastern Japan who adamantly opposed the policy decided in the early 1970s to stage a protest aimed at the government through the leader of one of the factions in the ruling party from the same district. It ended up being an abortive attempt because the police detected the governor's intent and took the necessary steps with the government. As another example, in an attempt to win rice farmers' consent and cooperation, instead of coercing them, a prefectural official in Kyūshū organized several community meetings through which, in the end, he succeeded in securing their approval. The merit of the policy – that is, the fact that farmers could receive subsidies without laboring to produce crops – became a decisive factor.

In 1978, the year of the second turning point, the target area for dormant rice paddies (where rice was not produced) was fixed, nominally, at 391,000 ha over 3 years. Partly due to the extraordinarily good harvest in the first year, 1978, and partly due to the overly optimistic outlook on the part of the government, it was discovered that, even if the 1979 goal were fulfilled, there would still be an oversupply of rice. To cope with this situation, guidance meetings were convened in various locations to request farmers to restrict rice production over and above the nominal goal. These attempts were called 'duck's underwater paddling' among the pertinent government officials, akin to a duck paddling hard underwater to swim along while looking calm above water. This 'underwater' guidance not only provoked criticism among local governments and agricultural cooperatives but also contributed to the rise of strong requests for the government to more forcibly set out explicit goals to make it easier to explain the policy to rice farmers. From the viewpoint of officials in the field, it was easier to request cooperation from rice farmers under a strengthened air of coercion than by stressing the virtue of self-restriction. In the end, the national government decided to maintain the announced goal as it was for 3 years. In 1980, the government set a new goal of 535,000 ha, or 1.4 times the original goal.

Rice paddy problem vs rice problem puzzles even the experts

The policy to curb rice production was a means to forgo rice surpluses by creating surplus rice paddies; that is, it was a method that replaced the rice surplus problem with a rice paddy surplus problem, so to speak. Personally, I hold the view that it was this replacement that began to make the rice problem hard to comprehend. For instance, today's rice policy includes the issue of wheat and soybeans to be planted in unused (or surplus) rice paddies. Similarly, surplus rice can mean both what is really a rice paddy area surplus as well as the rice surplus even after rice paddy area is adjusted. Rice policy and rice surplus are no longer taken at face value.

The perception among scholars is not uniform, either. We have entered an era when a mere reference to rice paddy agriculture, rice policy, or rice paddy management in the title of an academic paper can no longer tell us whether the paper deals with rice alone or includes non-rice crops produced in rice paddies as well as idle rice paddies, unless one reads its content.

Attention to rice paddies

To sum up, it has become necessary to differentiate between (1) rice, (2) rice paddies, and (3) the connection between rice and rice paddies since the second turning point in 1978. As if to correspond to this new trend, such agrologists as Yamane Ichiro and Yoshida Takehiko published books for general readers focusing on rice paddies around the second turning point. Yamane wrote *Nihon no Shizen to Nōgyō* (*Japan's Nature and Agriculture*) in 1974 (Yamane, 1974), while Yoshida published *Suiden Keishi ha Nōgyō wo Horobosu* (*Neglect of Rice Paddies Destroys Japan's Agriculture*) in 1978 (Yoshida, 1978). These authors paid special attention to the policy to curb rice production and, at the same time, stressed the value of rice paddies beyond that of merely being a place to grow rice.

In 1984, a group of journalists known as the Nōsei Journalist no Kai (Association of Agricultural Policy Journalists) convened a discussion titled 'Fate of the Rice Paddy' and featured the article 'Nihon ni totte suiden towa' (What do rice paddies mean to Japan?) in its quarterly, *Nihon Nōgyō no Ugoki* (*Trends in Japan's Agriculture*; Nōsei Journalist no Kai [Association of Agricultural Policy Journalists, 1984). Following the first turning point in 1967, which was the starting point of heightened interest in the Japanese people's relations with rice, the second turning point in 1978 became the point of origin of the heightened interest in the relations between the Japanese people and rice paddies.

Light and Shadow of Tasty Rice Production

Production of marketable rice

The quality of rice varies according to the breed and the soil and water where the rice is grown. Toward the end of the Tokugawa shogunate, a table of rice brand rankings was published; Kaga rice and Higo rice ranked top for western Japan and eastern Japan, respectively. Even under the Staple Food Control Act, which controlled rice across the board without any differentiation in quality, rice sellers did all kind of things to contrast their own rice with others. For instance, some rice sellers stressed the region where the rice was grown by using the name of the railway that transported the rice; other sellers marketed high-quality rice siphoned from rationed rice.

In the 1960s, consumer demand for tasty rice grew stronger, triggered by the increasing circulation of black market rice and the state's introduction of standards indicating the quality of particular brands of rice. This accelerated the full-fledged transformation of rice from goods to be shared to a commodity to be chosen by consumers. Today, which brand of rice is ranked Special A is a popular subject of discussion every year. Ranking the tastiness of rice was started in 1971 by the Japan Grain Inspection Association.

In 1978, the year of the second turning point, the Rice Price Deliberative Council, a government consultative committee, discussed whether disparities in quality should be considered when deciding the price of rice. The result was the differentiation among rice brands on the basis of quality that began in 1979. Brand used to be merely a standard for the convenience of distribution; as such, it did not affect the behaviors of rice farmers or consumers very much. Now the era has arrived when rice farmers choose the breed to plant, and consumers select the growing region and brand of their preference. While many breeds were grown around the first turning point, when the harvest stayed stable, the situation turned 180 degrees around the time of the second turning point. The Koshihikari brand was harvested most in 1979, and it remains the no. 1 brand today.

Also in 1978, a new national government ruling assigned a smaller area target for surplus rice paddies in prefectures producing tasty rice that was distributed at a high price as officially endorsed black market rice. This new rule prompted prefectures to start growing tasty rice breeds to reduce their assigned surplus rice paddy area as much as possible. Kumamoto prefecture, for instance, took advantage of the government's rice paddy usage reorganization measure to accelerate its drive to produce tasty rice. Aside from specifically assigning the breed of the officially endorsed black market rice and the areas where it could be grown, the prefectural government also promoted the planting of the Koshihikari brand rice in its early rice crop region. From 1978 to 1982, the amount of officially endorsed black market rice harvested in the prefecture increased by 9000 tons to 37,000 tons, and the share of rice harvested in Kumamoto as a whole increased from 12% to 21%.

With the introduction of officially endorsed black market rice, the rice rationing system by then had become ineffective. Even Minister of Agriculture, Forestry and Fisheries Watanabe Michio confessed at the House of Councillors' budget committee meeting in March 1979 that he himself did not possess the rice and grain passbooks. The Staple Food Control Act, which by that time was only nominal, was repeatedly reviewed from the late 1970s through to the 1980s. When the rice ration system was officially abolished, the officially endorsed black market rice found its way into law.

And this was only part of the story. It now became possible for rice farmers to give away their rice to individual consumers for free or to sell specially grown rice, including chemical-free rice, directly to consumers. Additionally, restrictions on the handling of officially endorsed black market rice were relaxed, diversifying connections among wholesalers as well as between wholesalers and retailers. As rice became liberated from the national government's stringent control through various regulations and monitoring, distributors and consumers came to recognize the significant difference in the quality and price of rice according to the growing region and brand.

While the officially endorsed black market rice continued to circulate, it was no match for a robust market mechanism devoid of government intervention. Markets dealing in unendorsed black market rice also emerged, including the Kanda Chūkai Shijō (Kanda Mediation Market) in Tokyo in 1979 and the Ōsaka Shōmai Shijō (Ōsaka Spot Goods Rice Market) in Osaka in 1987. In those days, even rice distributed through the formal channel was diverted to illegal channels. Although these illegal acts were repeatedly exposed by law enforcement authorities, brokers still acted as they pleased, unchallenged.

Birth of quality rice-producing regions

As of 2020, Niigata prefecture is the no. 1 rice producer in Japan followed by Hokkaidō. Yet both of them once had been far from being rice-growing regions. In fact, they were latecomers in terms of rice production.

Rice produced in Niigata prefecture in the 1960s had a bad reputation among wholesalers and retailers. Harvested in the muddy, swamp-prone Echigo plain, Niigata rice was not tasty. Although the Koshihikari breed had already been developed in 1956, its yield was relatively small. Under the system in which the government bought rice at a uniform price regardless of its quality, few rice farmers produced Koshihikari. Also, many rice farmers, agricultural associations, and local governments were reluctant to introduce officially endorsed black market rice or Koshihikari.

The Niigata Association of Economic and Agricultural Cooperatives, which dealt with officially endorsed black market rice, was an exception. It conducted an aggressive offensive to market Niigata-produced Koshihikari by introducing it to the Odakyu Department Store in Shinjuku and to other retailers in Tokyo, as well as to retailers in parts of western Japan where the Sasanishi brand, a rival produced in Miyagi prefecture, had not made inroads. Partly because western Japan had not developed rice as tasty as Koshihikari, this campaign succeeded in making Koshihikari of Niigata prefecture known all over Japan.

Hokkaidō, in contrast, faced hardship in its effort to harvest tasty rice. Hokkaidō rice was unsavory, but the region was too cold for Koshihikari. It was only natural, then, that Hokkaidō experienced an influx of rice produced outside of the prefecture when its people began to pursue tasty rice. Hokkaidō rice farmers were at the mercy of the national government for the sale of their rice. But Hokkaidō was a nuisance to the state as well as to other rice-growing prefectures. In fact, Hokkaidō sold such huge amounts of its unsavory rice that its rice was given the derogatory name Yakkaido rice: replacing the 'Hokkai' with 'Yakkai,' which means being a nuisance. As a part of the rice paddy usage reorganization measure, Hokkaidō was instructed by the national government to refrain from rice farming in 49.9% of its rice paddies.

Leaders of Hokkaidō's agricultural cooperatives in those days protested, saying, 'Rather than sitting back and doing nothing, we must take the offensive now' and 'There is no other way to survive than by competing on price alone' (Saeki, 1981). So Hokkaidō decided to sell its rice cheap and cover the deficit with funds contributed by all its rice farmers. It chose, therefore, to compete on price alone, deliberately absorbing the loss. This policy unexpectedly stimulated local farmers' consciousness about sellable rice, which in turn generated a drive among farmers to produce high-quality rice. Out of affection for and pride in Hokkaidō rice, new brands were developed over time. The Kirara 397 brand tasted so good that it drastically changed people's image of the taste of Hokkaidō rice, and the Nanatsuboshi and Yumepirika brands earned Hokkaidō recognition in people's minds as a producer of tasty rice.

Why do different kinds of rice paddy survive?

In the midst of a continuous rise in surplus rice paddy area, if farmers stopped growing rice in paddies that do not yield tasty rice as well as in paddies that are unproductive in terms of yield, starting with the worst ones, and worked efficiently in their other paddies, eventually there would remain in Japan only paddies that can produce tasty rice efficiently.

Although in Japan some rice paddies deemed unsuitable for crops were planted with trees or left barren, overall, rice paddies, including those with favorable conditions for rice growing, were randomly converted to other uses. This is why rice is still grown today in all kinds of paddies, big and small, on mountains and plains. The presence of part-time farmers and farmers harvesting rice for their own family consumption are a part of the cause for the situation, but there are other reasons as well.

First, producers whose rice used to be unpopular, including in Niigata prefecture and Hokkaidō, did not necessarily go into decline. In many cases, the policy to curb rice production was applied evenly to all rice farmers in each region with everyone sharing the pain equally, even though the area allotment rate might have been different from prefecture to prefecture. As such, the policy to curb rice production was not always applied to rice paddies in a sequence related to how unfavorable growing conditions were.

To begin with, rice paddies that produce tasty rice are not always the most efficient paddies. Even in well-known rice-producing regions there are not a small number of paddies located in low uplands where work efficiency is low. When a particular brand of rice is produced only in a limited range of regions, it takes on an additional scarcity value that, at times, means it can be traded at a high price.

On the other hand, not all rice paddies with favorable conditions, including location on a plain, good labor efficiency, and the capacity to produce tasty rice, have survived. Particularly during the period of high economic growth (1955–1973), and the national land development boom triggered by Prime Minister Tanaka Kakuei's Plan for Remodeling the Japanese Archipelago, land prices shot up, which generated tremendous pressure to convert farmland into industrial and other uses. Expectations that rice farmers would convert their rice paddies were especially high. This was good news to the Ministry of Finance, because reduced rice paddy area meant smaller subsidy payments for surplus rice paddies. The momentum toward the conversion of rice paddies was so great that in 1969, when the policy to curb rice production began, the then Secretary-General Tanaka Kakuei of the ruling Liberal Democratic Party proposed that laws for protection of farmland should be abolished and surplus rice paddy area should be converted to residential and industrial sites. Aside from the newly enacted City Planning Act, a provisional rule was put in place to make it easier to change the land category classification of rice paddies along national roads. This prompted the conversion of a number of rice paddies, including those that were suitable to rice farming, to residential and industrial sites.

Additionally, it should be pointed out that rice production is deeply connected with community relations, such as communal water use and communal works. As such, some rice farmers continued to produce rice regardless of efficiency or a lack thereof. Untended rice paddies can become breeding grounds for pests or shelters for wild boar and deer. Some rice farmers feel obliged to grow rice in order not to trouble their neighbors. On the other side of the same coin, some farmers gave up on growing rice because of unkept rice paddies adjacent to their own paddies.

As we've seen, reasons behind a farmer's decision to continue to grow rice or to stop growing rice were not uniform. Neither was the aftermath of those paddies where rice was no longer grown. Of the 373,400 ha of rice paddy area reduced between 1969 and 1977, 36% was converted to residential sites and 31% left untended; this is two-thirds of the total. The remaining one-third became farm roads and forestry roads or railways (12%), forestation sites (11%), or factory sites (10%).[1]

When rice paddy area and rice farming were reduced in a disorderly manner and when future prospects were utterly uncertain, it was a historical inevitability that an ideal picture of society and agriculture was envisioned for farmers to turn to.

The Envisioned Ideal and Setbacks

Conflict between criticism of agricultural policy and anxiety over food imports

The year of the second turning point, 1978, was also the year in which criticism of the government's agricultural policy began to be heard from the big business circle.

In the midst of a rice surplus and when the era of United States–Japan trade friction started with the unilateral cancellation by the U.S. government of the direct international convertibility of the U.S. dollar to gold, people and the government in Japan became more concerned about domestic agriculture because it had to compete with U.S. agriculture, which enjoyed much stronger international competitiveness. Later during the food crisis in 1973–1974, a council comprised of such business leaders as the presidents of Keidanren (the Japan Federation of Economic Organizations) and Keizai Dōyūkai (the Japan Association of Corporate Executives) announced that expanding food imports was the wrong decision. As the shock of the food crisis subsided, however, criticism from business leaders gushed out. They argued that the import of agricultural produce should be liberalized immediately, and that food and land prices had been pushed up by an overprotective agricultural policy, putting a strain on people's lives.

The Ministry of Agriculture, Forestry and Fisheries, which was reorganized from the Ministry of Agriculture and Forestry in 1978, became a target for thorough review of its expenditures under the Suzuki Zenkō cabinet formed in 1981. As a result, the ministry's budget continued to decline from ¥3701 billion in its peak year of 1982.

On the other hand, in an opinion poll on government policies conducted by the Prime Minister's Office in 1978, 70–80% of respondents approved of the government's effort to attain food self-sufficiency as much as possible, saying they would tolerate increased hardship if that was needed to attain a secure food supply. A Nippon Hōsō Kyōkai (Japan Broadcasting Corporation) TV program titled 'Day of Zero Food Imports' in March 1978 presented a provisional estimate that 30.51 million people, or a quarter of the entire population of Japan, would starve to death if imports of food were terminated. In November of the same year, the same TV station aired 'Stars and Stripes behind Dining Tables,' which covered the postwar history of rice and wheat centered around the U.S. food strategy. Both programs produced a lot of reaction among Japanese viewers.

Thus, the second turning point in 1978 was also the year that full-fledged contention began between advocates of promoting agricultural imports and people who urged caution.

Rice and the basic direction of agricultural policy in the 1980s

In the 1980s, Japan was urged to make a difficult judgment about the future direction of agriculture. The Council on Agricultural Policy, an advisory committee to the prime minister, submitted a report titled 'Basic Direction of Agricultural Policy in the 1980s.' The report's preamble said:

> Completing the catchup process with the west that Japan has pursued for 100 years since the Meiji period, we have become the no. 2 economic giant of the world next to the United States. We are about to enter the transition period toward the new era of a mature society. [translated from the Japanese original by this book's translator]

When Kōnosu Kenji, an agricultural ministry official, wrote these words, he had in mind Shiba Ryōtarō's novel, *Saka no Ue no Kumo* (*Clouds Above the Hill*). On the realization that the era during which Japan had aspired for the brilliant white cloud above the hill of the industrial west was over, the report pointed to the basic direction of Japan's agriculture at a turning point beyond which it had to pursue its own path.

Aside from discussing the policy to make rice production more efficient, the report explicitly singled out animal products as a rival to rice. While wheat had been an earlier archrival of rice, the shift in dietary trends among the Japanese from heavy reliance on cereals to increased consumption of animal products had become apparent by this time.

Recommending the Japanese-style diet

One of the messages from the 'Basic Direction of Agricultural Policy in the 1980s' was a new term: *Japanese-style diet*. The term referred to a dietary lifestyle based on rice as the staple, to be supplemented by vegetables, soybeans, fish, and seaweed and avoiding too much intake of animal fat, salt, and sugar. Some readers may immediately visualize the healthy Japanese meal consisting of rice, miso soup, and three dishes. As a matter of fact, on its webpage, the Ministry of Agriculture, Forestry and Fisheries advocates this healthy Japanese-style diet of the country with the world's highest longevity.

Among the various terminologies stemming from Japan's agricultural policies, none is more grossly misunderstood than the term Japanese-style diet. This is because the term was created to represent an ideal from the viewpoint of producers, as detected by the fact that the term was first advocated by the Ministry of Agriculture, Forestry and Fisheries ahead of the Ministry of Health, Labor and Welfare.

Behind the term was an aim on the part of the government to increase the proportion of domestically procured food as much as possible, thus minimizing dependence on imported food. The rice paddy usage reorganization measure launched in 1978 was a foreshadowing of this. In the cabinet meeting which approved the measure, it was argued that the measure was intended to actively promote a dietary life dependent on domestically obtained resources. In other words, a remedy for the persistent rice paddy area surplus and the promotion of a Japanese-style diet were pursued as a set.

Great role played by local community and agricultural cooperatives

The 'Basic Direction of Agricultural Policy in the 1980s' report also stressed the importance of the functions of agrarian villages as well as the roles of agricultural cooperatives and local communities. Recognizing the rise in the number of part-time farmers, the Council on Agricultural Policy, which authored the report, advocated fostering strong bonds among farmers. According to Hata Makoto, an official at the agriculture ministry, it was in 1967 that the ministry's view on part-time farmers began to turn from negative to positive. Also, triggered by a law enacted in 1965, a perception emerged that it was important to improve the living environments of depopulated regions and maintain a certain level of population in agrarian villages. The period between the first turning point in 1967 and the second in 1978 witnessed a spread of the view that part-time farmers were necessary for the stability and healthy development of local communities. This was also a period of heightened concern about the population balance throughout the whole country.

Inspired by the Second Comprehensive National Land Development Plan, which was adopted by the cabinet in 1977 (1 year before the second turning point), as well as the special measures taken by the agriculture ministry, new operations and systems began to emerge in this period. These enterprises reflected how conscious people were of the importance of proactive initiatives on the part of local communities, not only for rice policy but also for agricultural policy as a whole. Discussions in local communities and community works demonstrated this.

Regarding the sense of connectedness among farmers, around the same time local agricultural cooperatives intensified their engagement with farmers. Although agricultural cooperatives used to merely collect rice on behalf of the state, now they began to strengthen and widen their engagement with farm households; for example, by proposing a variety of measures such as growing various crops, sharing machinery and facilities, distributing produce, planning for regional agricultural development, and improving lifestyles in farming communities one after another in 1967, 1970, and 1979. Also, as the amount of officially endorsed black market rice increased, agricultural cooperatives, which controlled the majority of black market rice, strengthened their role as a forum to decide the rice price standard.

The operational outline for the rice paddy usage reorganization measure called for community-wide consultations to plan how rice paddies not used for growing rice would be utilized. Consensus on curbing rice production was promoted through regionwide organizational responses by farmers, including part-time farmers. Here mutual relations evolved between the state and the agricultural cooperatives in which the state used the cooperatives to assist in rolling out its policy to curb rice production, while the cooperatives worked together with the state by submitting requests on the policy from their own perspectives. The contributions of local agricultural cooperatives were so appreciated that in 1986 Japan Agricultural Cooperatives, also known as the JA Group, a national organization composed of regional agricultural cooperatives throughout Japan, was given status comparable to the local administration as one of the drivers of the policy to curb rice production.

Under the name of collective responsibility a stick was added to the policy to curb rice production in 1978. Even when individual farmers complied with the reduction goal, government subsidies would not be granted unless the goal for the community as a whole was fulfilled. The government used peer pressure to pursue its policy to curb rice production by taking advantage of farmers' desire not to trouble their farming neighbors and by nurturing local momentum to communally purchase machinery for growing soybeans in surplus rice paddies. Because water seeping from an adjacent rice paddy is detrimental to growing non-rice crops in a rice paddy, the national government offered a carrot in the form of a subsidy to help farmers work together to align surplus rice paddies next to each other. As part of the effort to encourage use of rice paddies not growing rice, a law was enacted in the 1980s to promote the lending and borrowing of farmland.

The allotment of rice paddy area under the rice paddy usage reorganization measure's strong air of coercion accompanied by carrots and sticks provided rice farmers with an opportunity to share hardship among themselves. This created a new sense of community among farmers. It was around the time of this second turning point in 1978, for instance, that discussions on community farming initiatives to be organized by farmers in a village became particularly active.

While the policy to curb rice production had started in 1969, its impact on agrarian villages became particularly profound after the launching of the rice paddy usage reorganization measure. Obata Yūjiro, the then governor of Akita prefecture, praised the measure as a revolution in rice production comparable to the land reform immediately after the Pacific War.

Ideal use of rice paddies

Around the time of the first turning point, only a few people, such as Tōhata Shirō of the Agricultural Policy Research Committee, had the foresight to discuss the issue of surplus rice paddy area. Around the time of the second turning point, however, experts one after another began to develop their own ideal scenarios for the issue, which was now a reality.

For instance, some argued that, instead of harvesting rice every year, it would be ideal to rotate rice with other crops at regular intervals, to fully utilize the potential of rice paddies. This method had been implemented in the mid-to-late modern period in Japan, particularly in the Nara and Osaka regions. The United States and Italy, where satisfactory water supply cannot be counted on, also use this method today. Rotating rice with other field crops regularly would not only save water and prevent and kill weeds, it would also dynamically change the farmland ecology, including the acidification, revitalization, and changes in particle composition of soil, the aggregation and erosion of soil nutrients, and the growth environments of microorganisms. Active utilization of this method was proposed as a way to realize agriculture that benefits from growing both rice and other non-rice crops.

According to crop scientist Kawata Shin'ichirō and his associates, it is estimated that this method was applied to 67,000 ha of rice paddy area in 1947 in Hokkaidō and to 20,000–25,000 ha of rice paddy area in other prefectures in 1950. In the majority of cases at the time, applying this method was a passive measure taken merely to respond to cold summer damages and draught. Now that rice paddy area was in constant excess, however, it was considered high time to popularize this method as a proactive agricultural measure. Nevertheless, in most cases rice farmers ended up rotating surplus rice paddies among themselves to, so to speak, share the pain. Three rice farmers agreeing, for instance, to a rotation wherein once every 3 years one of them would not grow rice in their surplus rice paddy area is an example of this. Moreover, it was pointed out that the merit of this method was not endless. Thus, to date, this method of rotating rice with other crops in the same paddy field has failed to establish itself nationwide as a new farming method.

An ideal to connect rice paddies with farm animals

It was once expected that changes in farming methods could bring about social changes as well. During this period, it was hotly argued that the time had finally come to realize the ideal of connecting rice paddies with farm animals. The 'time had finally come' was stressed because this ideal vision originating in the West had long been advocated in Japan.

In the late 18th to 19th century, during the Edo period in Japan, a revolutionary agricultural method was developed in the West in which farmers raised farm animals to produce livestock products, grew a combination of fruit and vegetables along with cereals for animal feed, and used animal feces and urine as fertilizer. These practices resulted in population growth, consolidation of a land ownership system, and, eventually, the development of capitalism. In Japan, in contrast, farm animals and rice paddies had a weak connection. Farmers used cows and horses to collect hay and plow rice paddies, but the animals were merely a supplement to human power.

Comparing Western and Japanese agriculture during the Meiji period, two schools of thought emerged. Some, such as German agricultural economist H. Maron, highly praised the Japanese way that did not require time or care to feed animals and dispose of their feces and urine. On the other hand, M. Fesca, a German agriculturist hired by the Meiji government, proposed that livestock raising should be introduced to Japan's agriculture. Fesca's view represented the common perception at the time that Japanese agriculture was behind the West.

Around the time of the second turning point in 1978, consumption of livestock products increased in Japan. Farmers were able to meet demand by reling heavily on imported animal feed. In the face of this situation, a proposal gained momentum to combine rice paddies and livestock animals to establish Japan's own agriculture. This would be accomplished by growing animal feed crops in surplus rice paddies and using livestock animals' feces and urine as fertilizer. Nevertheless, when cheap imported animal feed was available and chemical fertilizer was much easier to handle than animal feces and urine, this proposal had no chance of being realized. Particularly disadvantageous for the proposal was the large gap between the price of rice for animal feed and that of corn. Corn was a major commodity for U.S. agriculture, and it was difficult for the Japanese government to subside Japanese rice farmers to compete with U.S. corn.

A notion that there was no rice paddy surplus

A characteristic trend conspicuous among visions of ideal ways to handle the problem of surplus rice paddies in those days was the attempt to solve the problem simply by stressing that there was no rice paddy surplus.

According to calculations, when planting rice and other crops in rotation is adopted and rice production requires 1.5 million ha of rice paddy area, 3 million ha would be needed if rice is produced every other year, while 2.3 million ha would be required if rice is planted twice in 3 years. And the proposed method of growing wheat, soybeans, and animal feed, all of which relied on imports, in surplus rice paddies and growing rice in case of food shortages was also based on the notion that there was no rice paddy surplus when a possible food shortage was taken into consideration.

It has to be noted, however, that these methods assumed that non-rice crops can be grown in rice paddies. Thus, a goal was set in 1973 for rice paddies to be made usable as cultivated dry fields. Nevertheless, not all rice paddies were made usable.

Diversification of proposals on rice and rice paddies

It was around the time of the second turning point that values and ideals on rice production and the use of rice paddies diversified. On top of the notions and proposals just reviewed, the government, political parties, public officials, and scholars, as well as rice farmers themselves, submitted all kinds of proposals. This time can be described as a time of 'free and lively discussion by scholars and controversialists,' as once famously advocated by the Chinese Communist Party. The flood of ideas in this period is reflected in the book *80-nendai Nōsei to Shokuryō Mondai* (*Agricultural Policy in the 1980s and Food Problem*) edited by Kanpō Tsūshinsha (Kanpō, 1980). A collection of visions for Japanese agriculture proposed by the Ministry of Agriculture, Forestry and Fisheries, agricultural cooperatives, and various political parties, this book contained, in particular, versatile suggestions on the major issue of the time: rice and rice paddies.

The 3 years between the first turning point of 1967 and 1969, when rice production in Japan peaked, was a period when concern about the negative legacies of high economic growth and food safety, including apprehension about residual agricultural chemicals, began to heighten. Court cases on four major pollution-caused illnesses (Minamata disease, Niigata Minamata disease, Yokkaichi asthma, and Itai-itai disease) were among the causes for the heightened concern. A surge in concern for people's living environment occurred around the time of the second turning point, as exemplified, for instance, by establishment of the Japan Organic Agriculture Association in 1971.

In academia, Hasumi Takeyoshi and Suzuki Toshinori published *Yūki Nōgyō eno Michi* (*Road to Organic Agriculture*) in 1977 (Hasumi and Suzuki, 1977) to become pioneers in the field of organic agriculture in Japan. Shocked by the death of a rice farmer by agricultural chemical poisoning in July 1978, Une Yutaka, an agricultural extension worker in Fukuoka prefecture, began experimenting with restricted use of agricultural chemicals for growing rice. Another example of new rice farming methods proposed in those days was weeding by ducks, introduced by Okita Toshio around 1975, which was consolidated as the rice-duck farming method in 1985 by rice farmer Arata Seikō.

At the same time, the ideal use of soil and use of farmland based on its self-support capability began to be studied as a part of interdisciplinary studies encompassing agrology, fertilizer science, crop science, and agricultural economics. Outcomes of these studies were compiled in these publications: *Nippon no Chiryoku* (*Japan's Soil Capacity*) (Ogura and Ōuchi, 1976), *Shokuryō Jikyūryoku no Gijutsuteki Tembō* (*Technological Prospect of Food Self-sufficiency Capacity*) (Nōrin Gyōsei wo Kangaeru Kai, 1976), and *Nōrin Bunken Kaidai: Chiryoku Mondai* (*Annotated Bibliography of Agriculture and Forestry: Soil Capacity*) (Library of the Ministry of Agriculture, Forestry and Fisheries, 1978).

Farmers' Criticism and the Pressure Release Measure of Rice for Processing

Rice paddies not suitable for conversion to cultivated fields

Some rice paddies lay fallow with no crops planted. Agricultural economist Arahata Katsumi attributes it to three reasons in this time sequence:

1. when the subsidy for fallow rice paddies is excessively high;
2. when the rice price is so high that no other crop of the season is attractive enough, even when the subsidy is lowered; and
3. when a rice farmer retires from farming due to low profitability, a shift to another income source, or old age, irrespective of the policy to curb rice production.

While each of these reasons seems reasonable enough, I would like to add a fourth: the presence of rice paddies unsuitable for non-rice crops from the beginning.

One example of a crop grown in rice paddies filled with water is sagittaria (arrowhead), an ingredient for a traditional Japanese new year dish. Sagittaria is a hygrophyte that began to be produced widely, particularly in Hiroshima prefecture, as an alternative crop to rice. Triggered by the policy to curb rice production, farmers in the Saku region of Nagano prefecture began to culture edible crucian carp; their efforts still continue today.

Plants such as lotus root, water shield, Japanese artichoke, and rush, which in Japan is woven into tatami mats, as well as fish including carp and loach can be grown in rice paddies filled with water. However, the market for these kinds of produce was not wide enough to warrant the total conversion of all surplus rice paddies for this purpose. These crops, therefore, failed to constitute a main alternative use of surplus rice paddy area. Under the policy to curb rice production, it was only natural for rice paddies that could not produce anything else to become dormant.

Liberation from the obsession with rice

Some may find it surprising, but when the Ministry of Agriculture and Commerce was first established in 1881, it had no separate office to deal with rice. More surprisingly, the first volume of the ministry's PR newsletter criticized Japanese agriculture's exclusive devotion to rice farming in its prefatory note. Still, around 100 years later, in 1978, there remained farmers who continued to rely solely on rice growing. Although Japan has had a history of rice production increasing along with population growth, it is quite extraordinary to find rice farmers who grow field crops only when they are subsidized or farmers who would rather leave their rice paddies dormant when they cannot produce rice.

Hitherto, these phenomena have been attributed mainly to the obsession with rice among the Japanese people, particularly rice farmers, economic conditions, and government policies and agricultural cooperatives, which have been much criticized. In my view, however, at the bottom of these extraordinary phenomena is an uncomplicated fact that there are rice paddies that can only produce rice. Seen from that angle, it is not so much that the Japanese people, or Japanese rice farmers in particular, have been so obsessed with rice. Rather, they have been bound by the existence of rice paddies that are cultivated only for growing rice.

Even though liberation from obsession with rice was a slogan advocated in the Agricultural Basic Act introduced in Chapter 1, along with the development of self-supporting farmers, it failed to be accomplished. Agricultural economist Kanazawa Natsuki points out that one of the reasons behind the failure is the sheer abundance of rice paddies that are unsuitable from the beginning for conversion to cultivated dry fields. It was difficult to accomplish liberation from rice in these rice paddies.

Around the time when a departure from rice and community cohesiveness became important topics triggered by the policy to curb rice production, yet another movement emerged to encourage communal or regionwide production of non-rice products in rice paddies. The One Village One Product movement advocated by Governor Hiramatsu Morihiko of Ōita prefecture is representative. Under this new movement, whether the rice paddy in question was bound by its suitability for crops other than rice affected a community's choice of local specialty to grow and its development strategy.

Particularly well-known among those local specialties is the black soybean of Tamba Sasayama, Hyōgo prefecture. Tamba Sasayama is located at the bottom of a basin liable to suffer from a dearth of water. Farmers there who had given up on rice farming had been meagerly growing black soybeans in their rice paddies or the ridge between paddies. Inspired by the government's policy to curb rice production, local farmers began to grow black soybeans at full scale in their rice paddies so as to upgrade the produce to a local specialty. The result was the expansion of area planted with black soybeans to 96 ha in 1978. Tamba Sasayama, thus, succeeded in liberating itself from obsession with rice and creating a highly marketable local specialty by featuring the black soybean.

Departure from rice was easier for rice paddies having only a weak link with rice from the beginning. Saitō Matsutarō, a farmer of the Shōnai Sand Dunes, Yamagata prefecture, attempted to reinvigorate melon farming, which had been terminated during the Pacific War, along with growing rice in vinyl film paddy fields. Melon is a crop suitable to a field with good drainage, such as sand dunes. In time, vinyl film paddy fields disappeared from the Shōnai Sand Dunes, positioning the area to become a leading area for melon production in Japan.

In Hokota-machi (present day Hokota-shi), Ibaraki prefecture, where conversion of cultivated fields to rice paddies had become widespread in the 1960s, the number of rice farmers who gave up on rice and started producing watermelon and tomatoes increased rapidly. This was because it became easier to grow these crops in cultivated fields that used to be rice paddies. For example, it became easier to build a vinyl greenhouse on farmland that had been flattened for use as rice paddies. The conversion to rice paddies had also included irrigation, which solved the problem of water shortage, a traditional problem for farmers.

In Saga prefecture, one of the leading producers of soybeans in today's Japan, soybeans were grown in 454 ha of farmland in 1977, of which merely 135 ha were rice paddies. In 1978, soybeans were grown in 2593 ha of rice paddies, and the entire area for soybean production, including cultivated fields, also increased rapidly to 2790 ha. By adopting the method of rotating rice paddy use and non-rice paddy use and allocating rice paddies where rice was not to be grown among farmers in the region so as to share the pain of the regional restriction on rice production, agriculture that efficiently produced both rice and other non-rice crops began.

The departure of rice paddies from rice growing shocked not only farmers growing alternative crops in former rice paddies but also traditional growers of those crops. When a vegetable grower in Tsumanuma-cho (present-day Kumagaya-shi) of Saitama prefecture, one of the leading producers of Fukaya cibol, saw the cibol harvested from a former rice paddy, he was deeply impressed by its fresh greenness, saying, 'I had no idea cibol could be so green!' (Nakanishi, 1984). In the well-known producer regions, insufficient growth is bound to happen in a field that has been exhausted by a cibol harvest every year. While Fukaya cibol may have retained its brand prestige, its quality had degraded over time. Compared with well-used cibol fields, rice paddies, which had been used solely to grow rice, were virgin fields for cibol, allowing it to grow greener and more succulent.

Detailed account of rice for animal feed movement

Still, there were rice paddies that were unfit for conversion to cultivated dry fields. The existence of rice paddies that had no other use than growing rice led to the rise of a farmers' movement known as the Rice for Animal Feed movement. And in May 1981, experimental production of rice in surplus paddies for animal feed was officially approved. This movement aimed to uphold the sublime philosophy of connecting rice paddies and livestock animals as well as the goal of lessening Japanese animal husbandry's reliance on imported U.S. corn. It succeeded in its aim and, accordingly, rice for animal feed was produced experimentally in 367 cities/towns in 35 prefectures in 1982. The movement, however, proved to be short lived.

Some scholars attribute the movement's sudden termination to failure to develop rice breeds suitable for animal feed, while others denounced the movement itself as a temporary distraction launched by those who wanted to criticize the policy to curb rice production. It is my view, however, that this movement had evolved because there were groups of farmers united in their apprehension about rice paddy area that can only produce rice, and that, ironically, it ended for the same reason.

To illustrate my point, it should be pointed out that, by the time the Rice for Animal Feed movement came to an end, the policy to curb rice production had been converted to a policy to curb the production of rice for meals. As a part of this policy change, production of rice to be processed into miso, shoyu, and rice crackers came to be promoted in 1984. A subsidy of ¥70,000 per ton aimed to encourage farmers to produce 270,000 tons of rice for processing in 56,000 ha of rice paddies. Because stockpiles of surplus rice, which in those days was used for processing, were diminishing, the policy was intended to promote the production of rice exclusively for processing use.

Perhaps less impactful than the production of rice for animal feed was a new option for rice farmers to use surplus rice paddy area to grow rice plants for direct feeding to animals, instead of rice to be processed into animal feed. While feeding animals with rice plants in the pre-tasseling stage had been allowed until then, more often than not the pre-tasseling rice plant was simply disposed of. As a more practical alternative for feed, the whole crop silage (WCS) method (fermentation of stalk, leaf, and tassel clipped when they are still immature) began to be promoted; development of special machinery and breeds were started in Mie and Saitama prefectures.

The discovery that there was a way for surplus rice paddies to produce rice and rice plants, as these examples show, acted as a pressure reliever for the frustration that had been felt among farmers and regions with rice paddies unsuitable for non-rice crops. Seeing as the Rice for Animal Feed movement was driven by the sense of unity among those who were apprehensive about the future of rice paddies unsuitable for non-rice crops, this change in the policy to curb rice production quickly deprived the movement of its raison d'être.

Emergence of varieties of rice

These days it is no longer unusual to find bread made of rice flower in supermarkets. This would have been unthinkable 70 or so years ago when the rice substitute, 'manmade rice,' was made of wheat. Studies on the use of rice have a long history. Before the Pacific War when rice was imported smoothly from Taiwan and Korea, a research institute on use of rice was established within the Ministry of Agriculture and Forestry's Rice and Grain Management Bureau. The institute experimentally produced cosmetics and whisky from rice, but the project was scrapped due to the rice shortage. Around the second turning point year of 1978, however, not only conventional foods made from processed rice, such as sake, rice cakes, and Japanese sweets, but also technologies for new uses of rice, including for bread, noodles, and animal feed, came to be openly studied under the initiative of the agriculture ministry.

As the problem of surplus rice was transformed into a problem of surplus rice paddy area, a radical change of perspective occurred from how best to use surplus rice to how to produce rice suitable for many uses. Liberated from the assumption that rice is to be cooked and eaten, people generated a variety of ideas about rice breeds and processing methods. In the 1980s, interest in the exploration and utilization of genetic resources grew because of a Science and Technology Agency policy. And it was through a research project sponsored by the agricultural ministry in 1989–1994 that not only a high-yield rice species but also a number of new species of rice with a variety of nutritional compositions, appearance, and usages were developed.

Among the newly developed species were low-amylose rice, a highly gluey rice that does not harden when cooled; high-amylose rice, a non-sticky rice suitable for Chinese fried rice; and low-glutelin rice and hypoallergenic rice, which people with kidney disease and allergies can eat without after effects. Black rice and red rice, which had been grown despite their ill-treatment by the Staple Food Control Act, came to be highly regarded as healthy foods and underwent breed improvement. Fragrant rice, which had been loathed because of its distinct, popcorn-like fragrance, also came to be re-evaluated for its uniqueness.

Freed from the assumption that rice is grown solely so that people can eat it, research was launched not only regarding the aforementioned WCS, but also on ornamental rice, such as species with shapely leaves and tassels that can be made into dried flowers or cut flowers as well as those with extremely short stalks that can be made into bonsai.

The Year 1978 in Retrospect

At the second turning point, the issue of surplus rice was replaced by the issue of surplus rice paddy area. Rice had once been uniformly standardized and evaluated as military goods. After being liberated from these restrictions, rice became much diversified not only in its taste but also in its nutritional composition, usage, and appearance.

This newly set stage for surplus rice paddies inspired the emergence of a variety of ideal visions. Numerous new visions were presented and various movements were tried including the Rice for Animal Feed movement for rice paddies unsuitable for any other crop than rice. In retrospect, it seems there has been no other time in history when political parties, agricultural economists, farmers, and public officials fought out respective policy proposals and visions for ideal rice production related to land use and agricultural methods.

While new vision was brought to both rice and rice paddies, the future course of each was quite different. For rice, a solution to the problem of surpluses was found. But rice paddies were forced to bear the burden of the solution single-handedly. While rice started on a path of greater freedom, rice paddies were rejected as land to be cultivated exclusively for rice. This forced rice paddies where rice is not produced to explore new directions.

Some rice paddies succeeded in experimenting with new directions and departing from rice. But not all rice paddies were successful. It was difficult for rice paddies that can only produce rice to realize a new future vision. Although the government explored a solution to the surplus rice paddy area issue through providing assistance for production of rice for non-food uses, it failed to institute a formal legal system. This left the issue of the usage of surplus rice paddies in limbo.

These differences in future prospects between rice and rice paddies became more pronounced as domestic and overseas demand for freer trade of rice increased. It was the voice of the market that delivered the decisive blow. The pressure for free trade, which had been somebody else's problem earlier, steadily closed in on rice, eventually leading to the next turning point in 1993.

Note

1 Based on Ministry of Agriculture, Forestry and Fisheries data from various years, these figures for 1969–1977 are obtained by adding each year's record, excluding land damaged by natural calamity and that converted to cultivated fields. For the sake of simplicity, 'residential site and others' is included in

residential sites, 'farm and forestry road and others' in farm and forestry roads, and 'others' in abandoned untended land.

References

Bush, G.W. (July 27, 2001) President's remarks to the future farmers of America. Avai able at: https://georgewbush-whitehouse.archives.gov/news/releases/2001/07/text/20010727-2.html

Hasumi, T. and Suzuki, T. (1977) *Yūki Nōgyō eno Michi* [*Road to Organic Agriculture*]. Rakuyū Shōbō, Tokyo.

Higaki, T. and Terayama, Y. (1998) *Nōgyō-Nōson ni Mirai wa Aruka* [*Does Agriculture/Agrarian Village Have a Future?*]. Chikyusha, Tokyo.

Kanpō, T. (ed.) (1980) *80-nendai Nōsei to Shokuryō Mondai* [*Agricultural Policy in the 1980s and Food Problem*]. Nōgyō Seisaku Chōsakai, Tokyo.

Kawashima, T. (1967) *Nihonjin no Hō-Ishiki* [*Legal Consciousness of the Japanese People*]. Iwanami Shoten, Tokyo.

Library of the Ministry of Agriculture, Forestry and Fisheries (ed.) (1978) *Nōrin Bunken Kaidai: Chiryoku Mondai* [*Annotated Bibliography of Agriculture and Forestry: Soil Capacity*]. Nihon Nōgyō Bunko, Tokyo.

Mainichi Shimbun (1994) *Kome to Nihonjin* [*Rice and the Japanese*]. *Mainichi Shimbun*, October 28, 1994.

Ministry of Agriculture, Forestry and Fisheries (ed.) (various years) *Sakumotsu Tōkei* [*Statistical Survey on Crops*]. Ministry of Agriculture, Forestry and Fisheries, Tokyo.

Nakanishi, H. (1984) *Kome wo Ou* [*In Pursuit of Rice*]. Keisō Shobō, Tokyo.

Nōrin Gyōsei wo Kangaeru Kai (1976) *Shokuryō Jikyūryoku no Gijutsuteki Tembō* [*Technological Prospect of Food Self-sufficiency Capacity*]. Ochanomizu Shōbō, Tokyo.

Nōsei Journalist no Kai [Association of Agricultural Policy Journalists (1984) Nihon ni totte suiden towa [What do rice paddies mean to Japan?]. *Nihon Nōgyō No Ugoki* [*Trends in Japan's Agriculture*] 68.

Ogura, T. and Ōuchi, T. (eds) (1976) *Nippon no Chiryoku* [*Japan's Soil Capacity*]. Ochanomizu Shōbō, Tokyo.

Saeki, N. (1981) *Tokubetsu Jishu Ryūtsūmai noTenkai to Sono Mondaiten* [*Evolution of Special Independently Distributed Rice and Its Problems*]. Nōsei Chōsakai, Tokyo.

Yamane, I. (1974) *Nihon no Shizen to Nōgyō* [*Japan's Nature and Agriculture*]. Nōsan Gyoson Bunka Kyōkai, Tokyo.

Yoshida, T. (1978) *Suiden Keishi ha Nōgyō wo Horobosu* [*Neglect of Rice Paddies Destroys Japan's Agriculture*]. Nōsan Gyoson Bunka Kyōkai, Tokyo.

4 Internationalization of the Rice Problem: The Third Turning Point in 1993

Abstract

Chapter 4 covers the year 1993, when the Uruguay Round of the General Agreement on Tariffs and Trade was concluded, as the third turning point, and reviews the singularly Japanese history of rice in the midst of global expansion of the trade of agricultural produce. Externally, Japan's rice was made an exception in international negotiations as a non-trade concern, giving it special treatment distinct from other agricultural goods. Domestically, against the backdrop of a global trend to adopt market mechanisms in the trade of agricultural produce, rice distribution in Japan was converted from direct government control to a policy based principally on private transactions. This change was expected to prompt development of Japan's rice industry through free competition among rice farmers. Yet as the chapter points out, a national consensus on what to do with rice paddies where rice is not produced still remained unreached.

Tidal Wave of Trade Liberalization

Trade in agricultural produce

Punta del Este is a scenic beach resort in southeastern Uruguay. It is also where the Uruguay Round of the General Agreement on Tariffs and Trade (GATT) was launched. This multilateral negotiation, announced in 1986, continued for 7 long years. One result was a substantive agreement on rules of trade in agricultural produce in December 1993. This agreement triggered market opening for Japan's rice, which until then had been protected from the tide of trade liberalization. The setting of international rules on agricultural policy was also a turning point marking the end of the era when a country could freely decide its own goals and policies.

GATT was an international agreement signed by 23 countries in 1947 the overall purpose of which was to promote international trade; Japan became a signee in 1955. Behind this pursuit of free trade was regret that enclosure in colonies and exclusion of overseas products had become one of the causes of World War II. Tariffs either being imposed or reduced/eliminated through multilateral consultations without country bias brought with it not only economic merits but also pointed to the direction the world should take toward peace.

These ideals notwithstanding, unlike the talk on trade in industrial products, full-fledged discussion on trade of agricultural produce did not commence immediately. Dealing with agriculture in international negotiations was difficult because agriculture is directly involved in human life and its methodology varies greatly country by country. Particularly in the immediate post-World War II days, when many countries were busy restoring their own war-torn agriculture, rule-setting on trade was a secondary priority.

As a result of each country having restored its own agriculture freely, advanced countries generated an oversupply of agricultural produce in the 1960s. Some of those countries adopted a policy of requesting that their farmers abstain from farming. When the world experienced the food crisis in the 1970s and realized that food could be a third weapon after arms and oil, however, the situation took a sudden and complete turn. Because advanced countries pursued further expansion of agricultural production irrespective of existing oversupplies, clashes of interest surfaced among them.

DOI: 10.1079/9781800627093.0004

Public mood to call for international rule

Particularly fierce was the conflict between the European Community countries (present-day European Union countries) and the United States. Originally, European Community countries had relied on imports of agricultural produce from the United States and other countries. Now member countries followed a common policy of supporting the prices of European Community produce and refraining from importing low-price overseas produce. In time this allowed them to harvest enough produce to export. European Community countries began to export their oversupplied produce with subsidies.

This infuriated the United States. The European Community, which once had been a buyer of U.S. agricultural exports, now had turned into a rival in agricultural trade. Besides, expansion of agricultural exports from the European Community pushed down international prices of agricultural produce drastically. In response, the U.S. government increased subsidies for American agricultural production and increased its support for exports of produce. The end result of this rivalry was a trade war in which both sides continued to pour in government expenditures in order to win: pushing down the price of agricultural produce, which in turn forced governments to expand their expenditures even further. Needless to say, governments on both sides could not continue to subsidize their respective agricultures endlessly. Thus, it was only a matter of time before an international trade rule became imminent.

In 1986, when the Uruguay Round began, the ratio of agricultural budget to total agricultural output was 35% for the United States, while those of France and Germany were 58 and 27%, respectively. The same ratio for Japan, which succeeded in solving the rice surplus problem by curbing rice production, was 22%. This shows how large the agricultural expenditures of the U.S. and European Community governments were in those days.

Pressure of trade liberalization on rice

Before the third turning point in 1993, Japan experienced four waves of trade liberalization in agricultural produce. The first wave in 1960 resulted in import liberalization of 121 items of agricultural produce due to a cabinet meeting decision to promote liberalization of agricultural produce. In the second wave (1962–1964), soybeans, chicken meat, and onions were added to the list of liberalized items, leaving only restricted 72 items. Through the third wave (1970–1972), items to be liberalized were further expanded to include pork, apples, grapefruit, and compound feed. This wave also saw a raise in the import ceiling for beef and a reduction in the tariff on soybeans. When the fourth wave hit Japanese agriculture in 1978–1984 in the midst of U.S.–Japanese trade friction, Japanese produce came under a lot of pressure. Japan was forced to accept complete import liberalization and liberalization of beef, oranges, and orange juice importation.

All through these waves, rice was consistently treated as an exception. With the start of the Uruguay Round, which aimed at liberalization of all agricultural produce, however, foreign pressure on rice rapidly increased.

The rice war and Japan

Turning our attention to the world rice market, we see this situation: in the 1980s, rice prices all over the world fell because of the rice surplus that was partly due to expansion of rice production by the Green Revolution. Particularly in Thailand, the world's largest rice exporter in those days, inexpensive rice had been produced at the mercy of nature – through economizing labor and without any fertilizer or agrichemicals – in rural villages that were resilient to poverty. The Thai government had imposed an export tax on rice to secure revenue, but as the international price of rice went down, so did the export tax until it became zero in 1986. With the yoke of export tax lifted, Thailand began to export much more rice than before.

On the other hand, the United States, where rice production had developed as an export industry, was experiencing great difficulty in competition with major rice exporters. This was the so-called rice war. The country had suffered from a rice surplus problem since 1981, and a new subsidy was introduced in 1986 with the hope of supporting rice exports. The new subsidy drastically pushed up U.S. subsidies for rice from $24 million in 1981 to $961 million in 1986. With its massive fiscal expenditure as a weapon, the United States aggressively pursued dumping exports of rice, paying no heed to international criticism, to the extent that Thailand, too, resorted to dumping rice exports as a countermeasure.

In 1986, the U.S. rice stockpile reached a record high. Suffering from a huge financial burden to support rice in addition to trade friction with Japan, the United States began to attack Japan's rice market. On September 10, 1986, 10 days before the start of the Uruguay Round, American rice millers filed a complaint with the United States Trade Representative (USTR) demanding the opening of Japan's rice market. As a result of bilateral negotiations with Japan, the United States later withdrew the rice millers' complaint in return for inclusion of Japan's rice in the agenda of the Uruguay Round.

Full self-sufficiency in rice policy triggered by contamination of rice for processing

In the 100 years since the start of the Meiji period (1868–1912) until the first turning point year in 1967, rice imports were nothing unusual for Japan. It was in the 1980s that the public mood rapidly shifted to demanding domestic rice, particularly for day-to-day meals. A major catalyst of the change were the 4 years of poor harvests from 1980 (2 years after the second turning point). Rice for processing was gravely affected by these bad harvests. Because rice for processing in those days had a role to make up for shortages or oversupplies of rice for meals, lean crops of rice for meals directly took a toll on rice for processing. In 1984, when the Japanese government decided to use stockpile rice for processing, it found that the stored rice was contaminated with bromine, upon which it hastily reached an agreement with the South Korean government in June that year to return part of the rice the Japanese government had loaned to Korea.

The government's decision to easily resort to overseas rice while restricting rice production in Japan drew fire from Japanese farmers and agricultural cooperatives (members of the Japan Agricultural Cooperatives or JA Group, a cooperative society of farmers) even though South Korea only returned the rice it had borrowed earlier. Facing domestic criticism over easy reliance on overseas rice, the Japanese government made a rule that rice returned from Korea would be limited to use for processing. Within a month of the bilateral agreement, the two houses of the Diet resolved, for the first time, that full self-sufficiency of rice would be firmly maintained lest Japan should rely on foreign rice again.

More than 50% of the rice that was returned from South Korea in the year after the Diet resolution was found to be insect-infested. Moreover, some rice dealers were discovered illegally diverting some of the rice from Korea to meal purposes. The Japanese people's trust in the government's rice policy was so damaged that Ishikawa Hiroshi, who had just been appointed to director-general of the Food Agency, had to travel around Japan apologizing. But criticism and apprehension regarding rice imports became louder.

Cross-pressured by domestic policy and international negotiation

Subsequently, in 1988, pressured by a sense of urgency in the face of the wave of liberalization for the rice trade, both houses of the Diet unanimously adopted a resolution to oppose the liberalization of rice. This resolution once again confirmed the government's policy to achieve self-sufficiency in rice. Nevertheless, the Japanese people's wishes were by no means unanimous. In this policy for full self-sufficiency in rice, there was definitely an aspect of political performance vis-à-vis domestic and overseas audiences in preparation for election campaigns and international negotiations.

The bureaucrats of the Ministries of Foreign Affairs and Agriculture, Forestry and Fisheries were the ones cross-pressured by domestic policy and international negotiations. Although officials of these ministries attempted to make rice an exception of trade liberalization, these bureaucrats, particularly those directly involved in the negotiations, were well aware that it would be difficult to make Japan's rice an eternal exception in the midst of the global tide for trade in agricultural produce. When the ruling Liberal Democratic Party next attempted to pass a Diet resolution on full self-sufficiency in rice, these two ministries obstructed the move. At the time, Ōmori Masasuke of the Cabinet Legislation Bureau had run around trying to put out fires by explaining to Diet members that a Diet resolution would be useless because it had no binding legal power.

As the Uruguay Round neared conclusion, farmers aboard 500 light trucks took to the streets in protest against rice importation (*The Mainichi Shimbun*, December 11, 1993). (Used with permission from Mainichi Shimbun)

Domestic criticism on treatment of rice and arguments for partial opening

The government's effort to protect rice from the wave of trade liberalization was also criticized from within Japan. Most of all, the government pushed up the rice price too much. Its policy of making up for the income disparity widened during the period of high economic growth by raising the price of rice ended up with the domestic rice price being ten times higher than the world market price.

Procrastination on the liberalization of agricultural trade was also a drag on exports of industrial products. When Japanese export industries were hit hard by the 1985 Plaza Accord that led to rapid appreciation of the yen, voices demanding liberalization for agricultural produce became much louder than before. As liberalization of agricultural produce other than rice advanced, the special treatment enjoyed by rice became much more conspicuous. It was only natural, therefore, that demands for a reduction in the rice price and criticism of the government policy to reject rice importation became fiercer.

Around the time international negotiations began, two things happened in Japan. In 1985, Prime Minister Nakasone Yasuhiro's cabinet came out with a policy to actively promote a free trade regime. There followed in 1986 a report submitted by a private consultative body of Prime Minister Nakasone recommending further internationalization of Japan's agriculture. This report functioned as a testing ground for appointees to ministers of the Nakasone cabinet. Triggered by this report, criticism of agriculture and rice from the industrial sector grew all the more strong, resulting in reduction of the government's purchase price of rice in 1987. Also, the argument for a partial opening of rice trade with a fixed import ceiling became vocal among both the ruling Liberal Democratic Party and opposition parties.

Perception gap over rice

If we focus only on rice itself as a commodity, it may appear that the debate on whether rice trade should be liberalized or not – which divided public opinion in Japan in two – was an ideological rivalry between those advocates of alignment with the international trend, seeing opening the rice market as a way to treat rice more efficiently, on the one hand, and those who adamantly opposed rice imports to protect the vested interest in domestic rice, on the other. In actuality, however, whether to agree with rice liberalization or not was also affected by the meaning and value attached to rice. This meant that the two sides were arguing on different planes.

Even when the two sides shared a common perception, such as the notion that the Japanese love to eat rice, for instance, opinions diverged. Those who recognized rice as a mere commodity considered it the government's responsibility to liberalize rice and provide the people, who loved rice, with inexpensive foreign rice. Those who were strongly conscious of the food security aspect of rice argued that, no matter what, it was the government's duty to protect domestic rice for the benefit of the rice-loving Japanese people. In those days, thus, arguments over the rice issue involved the meaning and value of rice beyond a mere commodity; people spoke of, for example, the risk of reliance on foreign countries for the nation's staple food and influences on rice paddies as well as agrarian villages.

What merits special attention was that, by that time, the rice surplus problem had already been transformed into the surplus rice paddy area problem. As imported rice rushed into the Japanese market, the amount of rice needed to be produced domestically was lowered, making the surplus rice paddy area problem all the more serious. The struggle and soul searching that accompanied trade liberalization led to clarification of the perception gap among the Japanese over rice and rice paddies. In the end, this helped people find common ground, instead of deciding on an extreme position of either full acceptance or complete rejection of trade liberalization.

The Day the Rice Price Drop Became Destined

Non-trade concerns and GATT's conclusion

There is no doubt that trade enriches our life. It does not mean, however, that everyone wholeheartedly welcomes trade with no limits whatsoever. One reason behind reservations about trade is the risk that weak home industries might be damaged from trade. But that is not all. There is an internationally shared common perception that consideration must be paid to influences of trade on factors originally unrelated to trade.

As a country that relies on imports of food, Japan caused concern around the world with its December 1987 announcement that a country should be allowed to limit imports of its fundamental food. Although this position was in direct conflict with the U.S. and European Community argument for free trade without exception, by the time Japan introduced this reasoning into international negotiations in October 1988 it was no longer an odd notion.

It should be noted, however, that 'fundamental food' is a highly vague concept that harbors the danger of opening up a Pandora's box should each country start justifying its import controls by citing exceptions for fundamental food. In order to prevent this, an interim GATT agreement decided that a wide range of factors which were not directly related to trade but which merited attention, including the food security issue, should be labelled 'non-trade concerns.' These included, aside from food security, the environment, protection of plants and animals, public health and safety, and cultural diversity, among others. The target of non-trade concerns was not confined to agriculture and agricultural produce.

The role of rice paddies and rice production as a means to prevent flooding was also one of the non-trade concerns. According to a Mitsubishi Research Institute study commissioned by the Ministry of Agriculture, Forestry and Fisheries in 1991, the role played by rice paddies and rice production, aside from enabling the harvesting of a commodity, was estimated to be worth ¥4.7–11.9 trillion per year. Introduction of the concept of non-trade concerns signified the birth of a channel to evaluate the role of agriculture other than in terms of production. Discovery of a point of compromise on trade liberalization between aggressive advocates and those taking a hesitant, cautious position played a decisively important role in promoting a concession beyond the differences in position of each participating country.

Subsequently, the World Trade Organization (WTO) was established to succeed GATT, and trade rules came to be applied more stringently to create a new international order.

United States–Japan secret consultation in the summer of 1993

Let us go back to 1 year before the substantive agreement at the Uruguay Round. In November 1992, when settlement of the European Community–United States confrontation finally appeared reachable, negotiation on Japan's rice was the last unsettled item. At this point, Japan found itself isolated: by vigorously defending its rice, Japan could jeopardize the entire negotiations that participating countries had achieved. In the face of this situation, Kyōya Akio, vice-minister of agriculture, forestry and fisheries, was assigned to attend to domestic political coordination, while Shiwaku Jirō, vice-minister for international affairs, was put in charge of international negotiation. Although both of them were scheduled to retire from the government at the end of March 1993, their retirement was postponed until the conclusion of the negotiation. They were engaged in the international negotiation as their last service in office.

Initially, bilateral negotiations were on parallel tracks. But when Bill Clinton succeeded George W. Bush, a fundamental believer in free trade, as the U.S. president in January 1993, the U.S. side changed its policy to looking for a breakthrough. On July 19, 1993, Charles J. O'Mara, Special Negotiator for Agriculture to the U.S Trade Representative, approached Shiwaku with a proposition to look for a breakthrough in rice negotiations using the concept of non-trade concerns. This triggered advancement of the bilateral negotiations toward conclusion.

Subsequently, United States–Japan negotiations on rice were designated confidential consultations, excluding the Ministry of Foreign Affairs and the U.S. Department of State, to be conducted between O'Mara and the acting associate administrator of the Foreign Agriculture Service of the U.S Department of Agriculture and Shiwaku and Kumazawa Hideaki, director-general, International Affairs Department, from the Japanese Ministry of Agriculture, Forestry and Fisheries. Around the same time, the ruling Liberal Democratic Party went out of power and gave way to the Hosokawa Morihiro cabinet, a coalition government comprised of all seven opposition parties excluding the Japan Communist Party. Shiwaku and his fellow negotiators, however, steadily advanced negotiations with the U.S. side behind closed doors without obtaining approval from the coalition government parties, including the Japan Socialist Party.

The cold summer of 1993

In the summer of 1993, when the behind-closed-doors negotiations were being conducted between Japan and the United States, all of Japan except for Okinawa was hit by unusually cold weather. Meteorologists could not even declare the end of the annual rainy period. Attributed to the massive volcanic eruption of Mt Pinatubo of the Philippines, this cold summer brought a record poor harvest to the archipelago. Such descriptions as 'the rice riots of Heisei' (the rice riots of 1993) and 'the year Thai rice was sold at supermarkets' would surely ring a bell for any Japanese over 40 years old. Overall, that year's harvest was only 74% that of an average year. But in particularly hard-hit northern Japan, Aomori and Iwate prefectures experienced 72% and 70% drops respectively.

As we will see later in this section, selling of rice on the black market was rampant in those days, making it so difficult for the government to procure rice that it was able to acquire only 20,000 tons of rice in 1993. The government had no other option than to hurriedly hunt for rice produced in Thailand, China, and the United States. But the foreign rice was unpopular among Japanese families and rice dealers. Families found the foreign rice unpalatable, and, moreover, rice dealers found it difficult to obtain in bulk. Thus, emergency imports of rice failed to heighten people's concern about full self-sufficiency in rice. More than that, the imports aroused suspicion about the feasibility of the government's policy to block the import of rice. As a result, the movement to protect rice from trade liberalization ended up losing momentum.

In the midst of a rice shortage, consumers rush to an electric appliance store selling unendorsed black market rice cheaply at the risk of running a deficit. Miyaji Toshio of Jōnan Denki, who made the decision to sell rice cheaply, burst onto the scene as the sudden talk of the town. (*The Asahi Shimbun*, March 27, 1994). (Used with permission from Asahi Shimbun)

Conclusion of negotiations

When all the other participating countries at GATT had reached an agreement on agricultural produce, the chairman of the negotiating group, of which Japan was a member, presented a mediation plan to induce concession from Japan. Seeing as this plan included a rule allowing rice to be an exception, Japan decided to accept this plan on December 14, 1993, immediately before the negotiation deadline. As a result, Japan's rice became subject to international trade rules; for Japan, this meant the partial liberalization of rice imports. Under these new rules, in return for a 6 year grace period on tariffication, Japan was requested to import a minimum amount of rice even though Japan maintained full self-sufficiency in rice. The minimum amount of rice Japan was obliged to provide a minimum opportunity for imports. Rice imported under this international agreement is called 'minimum access rice.'

This special rule applicable only to Japanese rice so as to make it an exception was improvised at the last minute by officials directly engaged in the negotiations. One of the conditions attached to this special treatment was for Japan's rice to have an element of non-trade concerns including food security and environmental protection. It shows that the non-trade concerns played a decisive role as a lifeline to let Japan's rice narrowly escape from tariffication of rice without exception. It also presented a point of compromise between total acceptance of rice liberalization and complete rejection.

Additionally, the Japanese negotiators also succeeded in inserting in the preamble of the agreement a sentence saying that consideration of non-trade concerns was one backdrop to the conclusion of the agreement. It was also stated explicitly, upon Japan's proposal, that non-trade concerns would be an important agenda point in the renegotiation scheduled to commence in 2000.

These last-minute proposals from Japan were accepted partly thanks to the United States, which prioritized conclusion of the negotiations over anything else, as well as persuasions from the GATT secretariat. Thus, while the concept of non-trade concerns was a fruit of Japan's negotiation, it remained obscure lest each country accept it too rigorously as a trade rule. As of today, there still exist differences in the degree of conviction for this concept among countries in the world.

When after the conclusion of the negotiations Japan found that it could control the amount of imported rice better by tariffication than by making it an exception, it adopted tariffication of rice in 1999. Many have been critical, asserting that, had Japan adopted tariffication from the beginning, it could have restricted the amount of subsequent rice imports.

My view, however, is that it was more significant that treatment of rice as an exception was created as a fait accompli, by taking into consideration such non-trade concerns as food security and environmental protection. From the viewpoint of the historical positioning of Japan's rice production and agriculture, I find this more compelling than a debate on the volume of rice imports. As a matter of fact, having made the world recognize the argument for non-trade concerns and having successfully made exporting countries compromise so as to conclude negotiations, Japan became increasingly interested in those non-trade concerns.

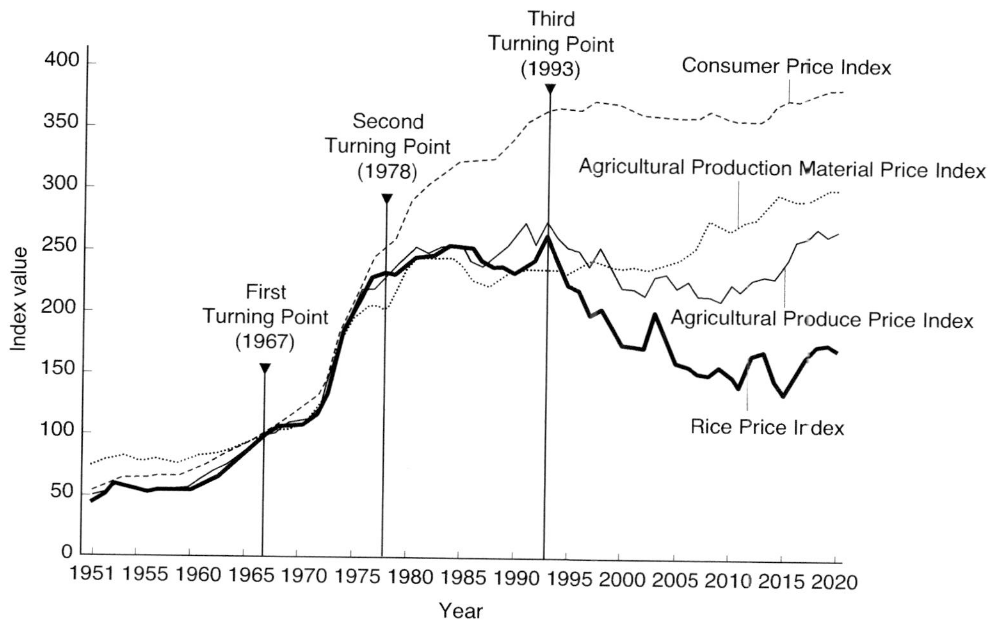

Fig. 4.1. Changes in Consumer Price Index, Index of Agricultural Production Material Price, Agricultural Produce Price Index, and Rice Price Index. *Source*: based on the Ministry of Internal Affairs and Communications' *Shōhisha Bukka Shisū* [*Consumer Price Index*] and the Ministry of Agriculture, Forestry and Fisheries' *Nōgyō Bukka Tōkei Chōsa* [*Statistic Survey on Agricultural Produce Prices*]. *Note*: all figures are index numbers for the whole of Japan using 1967 as the base year. Changes in Consumer Price Index exclude owners' equivalent rent.

Declining price of agricultural produce

Being produced by a large number of farmers, agricultural produce is liable to be bought at cheaper prices than industrial products that are manufactured by a limited number of companies. For this reason, prices of agricultural produce tend to be disadvantageous for the seller compared to general commodities and such agricultural production materials as fertilizer.

Let us look at the trends in the price of agricultural produce using the first turning point year of 1967 as a benchmark (see Fig. 4.1). From 1967 to 1978, the second turning point year, both the general price level (Consumer Price Index) and the price indicies of agricultural production material and agricultural produce went up together in a similar way. From around the second turning point, the two began to diverge, making it increasingly more difficult than before to purchase industrial products with the money earned from the sale of agricultural produce. From around the third turning point in 1993, price indices of agricultural production material and agricultural produce also began to diverge. Particularly noteworthy was the big drop in the price of rice. Thus, prices of agricultural production material, including agricultural machinery, fertilizers, and pesticides, produced by manufacturers came to be felt to be relatively expensive.

One of the causes of this phenomenon was that all agricultural produce was now obliged to comply with the trade rules agreed to at the Uruguay Round of the GATT. This situation promoted the import of inexpensive produce produced overseas. Although rice was given an exceptional position at the Round, rice prices had declined because the newly made international rule prohibited the Japanese government from supporting the rice price, which had been set at a high level to start with.

What International Rules Brought to Rice Paddies

Conditions for agricultural policy stipulated by international rules

Common rules for agricultural policy were also decided at the Uruguay Round. Subsidies and export subsidies to promote the expansion of production, for instance, now were reduced because they might become a cause to upset trade. The era when each country was free to decide its own agricultural policy gave way to a new era in which even domestic policy was restricted by international rules.

The most typical change was emergence of an international order that discouraged policies which protected a country's own agriculture through price supports. The new order aimed at fairness by leaving prices to the market mechanism. It was agreed, however, that governments were allowed to support farmers who could become destitute when the price of agricultural produce went down by offering subsidies, provided such would not encourage the expansion of production. While this stipulation had already been adopted as a new principle of agriculture policy in a separate international negotiation in 1987, it was confirmed as a fully endorsed rule by the conclusion of the Uruguay Round.

When a government protected farmers by setting the price of agricultural produce high, farmers were supported by the money that consumers paid. In the new international trend of agricultural policy, the general inclination was for the mechanism to support agriculture by shifting the burden on consumers to taxpayers.

Homework for each country to renovate agricultural policy

Each country was now expected to reform its own agricultural policy in response to the new international rules. Particularly noteworthy here was the prompt response by the European Community, which had decided on its first agricultural policy reform in 1992, prior to conclusion of the Uruguay Round. Accordingly, the European Community adopted a policy for 1993–1995 to both let the market decide the price of agricultural produce with little governmental intervention and, at the same time, support farmers' income directly with government subsidies. The price of cereals, for instance, was reduced by as much as 30% in 3 years, for which farmers were compensated with a guaranteed income to make up for the drop in price. In compliance with international rules, the European Community government supported farmers' income through production adjustment, which would not directly promote production expansion, as well as subsidized farmers who engaged in environmentally conscious agriculture.

The United States, for its part, thoroughly reviewed its agricultural policy in 1996. The U.S. government decided to suspend regulations on agricultural production and let the market mechanism decide prices. In order to compensate for the loss incurred by farmers, the government decided to provide subsidies on the basis of past records.

As exemplified by the European Community and the United States, each country was requested to reform its own agricultural policy in accordance with the newly made international rules. Subsequently, countries have continued to update their policies to date.

Settlement of international negotiation

In contrast to other countries, Japan failed to respond promptly to the new international agreement. Hashihara Hisao, who assisted the international negotiation as assistant director of the planning division of the Food Agency, reminisced in later days that he learned the hard way that domestic reform should have preceded market opening instead of vice versa.

Japan's having consistently pursued self-sufficiency in rice meant that when the government decided to accept the proposed agreement at the Uruguay Round, before anything else, it had to obtain the Japanese people's understanding. Thanks to leaks by the U.S. media, the fact that rice could no longer be protected became widely known among the citizenry, but the government persistently denied the rumor.

On December 14, 1993, Prime Minister Hosokawa Morihiro's cabinet decided at its extraordinary cabinet meeting to accept the proposed agreement. The prime minister's press conference immediately followed just before dawn. Both the Ministry of Agriculture, Forestry and Fisheries and the Ministry of Foreign Affairs drafted versions of the prime minister's announcement and the actual agreement at the cabinet meeting, but it was the agricultural ministry's draft that was adopted. Minagawa Yoshitsugu, an official of the Minister's Secretariat Planning Division, Ministry of Agriculture, Forestry and Fisheries, authored the two draft documents after reviewing Emperor Shōwa's August 15, 1945, radio broadcast – known in Japan as the Jewel Voice Broadcast – proclaiming Japan's surrender, which his superior, Watanabe Yoshiaki, had handed to him to study. Minagawa succeeded in drafting the documents not from the viewpoint of Japan's having been defeated in international negotiations, but rather as a positive manifesto for the future. Minagawa's draft envisioned agriculture in Japan being an attractive and internationally harmonious industry, while at the same time admitting that the decision to accept the Uruguay Round's proposed agreement was, nevertheless, a last-minute, agonizing, and heart-rending decision for Japan.

Prime Minister Hosokawa Morihiro at the televised press conference convened immediately after the decision on partial opening of Japan's rice market. With the help of a teleprompter, he talks straight to the Japanese people, never lowering his eyes, on the conclusion of the Uruguay Round negotiations. Hosokawa became the first Japanese political leader who used this device (*The Asahi Shimbun*, December 14, 1993). (Used with permission from Asahi Shimbun)

Mitigation of the pain inflicted by international negotiations

Three days after the press conference, the Hosokawa cabinet decided not to tighten the policy to curb rice production any further, even when the minimum access rice was imported. This cabinet decision, most of all, eliminated the anxiety on the part of Japanese farmers that the imported rice would accelerate the problem of surplus rice paddy area. The decision also reassured people that when the minimum access rice was imported, it would be treated separately from the domestic rice market to prevent it from affecting domestic rice production.

The next year, in 1994, it was decided that a total of ¥6.1 trillion over 6 years would be spent for the purpose of responding to the consequences of the Uruguay Round. The government was not allowed to issue deficit-financing bonds for this purpose. Measures in response to the Uruguay Round were financed by construction bonds, which were rigidly confined in usage to such items as public works/construction, disbursements to relevant organizations, and loans to farmers. Also, rice that was exempted from tariffication was not eligible for this government support. With this government support, not only farmland and farm roads were improved, but also a number of vegetable stands and hands-on agricultural farms were constructed. In addition, hot spa facilities were constructed in 26 locations all over Japan by 1999. As these examples show, the government's measures responding to the consequences of the Uruguay Round had a strong aspect of stopgap measures to appease dissatisfaction of farmers and rural areas rather than being a well-thought-through strategic measure to prepare the internationalization of Japan's agriculture.

New problems facing rice and rice paddies

Bound by new international rules, Japan's agricultural policy underwent a fundamental reform of its legal system. Meddling in rice prices, in particular, brought a number of new problems. To interfere with prices automatically means to interfere with markets and distribution itself. More than that, it became necessary to review the policy to curb rice production, which had contributed to support for rice prices. In order for the government to back out of this support, a wide range of responses was required, from distribution of rice to use of rice paddies.

High Ideal Created as a Reactive Approach

Philosophy of the new basic law

In July 1999, the Agricultural Basic Act, also known as the constitution of the agricultural sector, was abolished; in its place, the Basic Act on Food, Agriculture and Rural Areas was enacted. This new basic law was made to set the direction for new food, agricultural, and rural areas proposed by the Ministry of Agriculture, Forestry and Fisheries during the Uruguay Round negotiations.

The basic law set out two principal philosophies regarding securing of a stable food supply and fulfillment of the multifunctional roles agriculture is endowed with. The term *food security* was adopted formally in law for the first time. Also, realization of the multiple functions of agriculture was explicitly stressed in the law. In order to accomplish these goals, the law aimed at promoting sustainable development of agriculture and rural areas. The multiple functions of agriculture include, according to the Science Council of Japan: (1) conservation of national land, (2) water source cultivation, (3) preservation of the natural environment, (4) formation of a good landscape, and (5) maintenance of cultural traditions. Multifunctional roles refer to the variety of functions generated from rural areas and agriculture, in addition to agriculture's conventional role as a primary food supplier. All of these, together with the philosophy of food security, correspond to the non-trade concerns highlighted during the Uruguay Round negotiations. The law was enacted before international negotiations over non-trade concerns commenced in 2000, making the Japanese government prepared for those negotiations.

Nevertheless, concrete legal systems to help realize these philosophies were inadequate. Comparing the basic law to mountain climbing, Takagi Yūki, the then vice-minister of agriculture, forestry and fisheries, reminisced in later days that, although the new law succeeded in envisioning the philosophical summit, it failed to outline the climbing route to the summit.

The Basic Act on Food, Agriculture and Rural Areas was enacted ahead of time in anticipation of the international negotiations scheduled to begin in 2000. Because the law was, so to speak, a patchwork to prepare for the negotiations on non-trade concerns, it was the product of a reactive response. We see evidence of this by the fact that, of the four purposes of the law – Securing a Stable Food Supply (Article 2), Fulfillment of Multiple Functions (Article 3), Sustainable Development of Agriculture (Article 4), and Development of Rural Areas (Article 5) – Article 3 was devoid of corresponding basic measures to be taken (see Fig. 4.2).

The policy to curb rice production was defined as a measure for 'price formation and stabilization of farm management' in the overall pursuance of 'sustainable development of agriculture,' but its relation to the fulfillment of multiple functions was not explicitly indicated. While absence of a detailed legal system acting as a bind has the merit of allowing flexible responses in international conferences, unfortunately the negotiations did not go as smoothly as Japan had hoped.

Japan aspires to coexistence of various forms of agriculture

In the proposal submitted to the WTO in 2000, the Japanese government stressed that the 21st century should become an era when various forms of agriculture in various countries and regions could coexist. It proudly proclaimed that

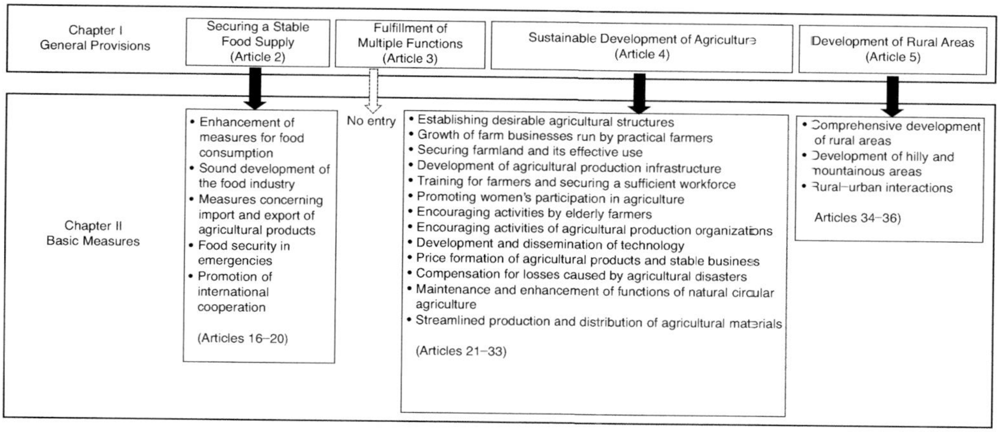

Fig. 4.2. General provisions of the Basic Act on Food, Agriculture and Rural Areas and Basic Measures.

the entire world should pursue securement of a stable food supply and fulfillment of the multiple functions of agriculture. It was a proposal to promote agriculture's multiple functions and secure food's stable supply through mutual recognition of various forms of agriculture differentiated from each other by natural conditions and histories. It was Japan's argument that agriculture must be protected to let agriculture fulfill its multiple functions. Japan prepared itself for the international negotiations that began in 2000 by making a domestic policy that advocated the philosophies of non-trade concerns and earnestly reducing those subsidies that should be pared down.

Stagnated negotiations

Fellow countries sympathizing with Japan's call for coexistence of various forms of agriculture proved to be few in number, however. This was not so much because Japan negotiated poorly, but because each country was having a hard time finding allies. At the Uruguay Round beginning in 1986, there was an easily detectable axis of contention, such as the confrontation between the United States and the European Community. As time went by, however, the world situation changed drastically; the axis of contention and conflict of interest had become markedly more complicated.

Not only did the number of countries participating in the negotiations increase from 123 to 151, but the confrontation between advanced countries and developing countries also became much more pronounced. More than that, with the remarkable rise of China, India, and Brazil among the middle-income countries, the axis of confrontation was no longer a simple one of high-income countries vs low- and middle-income countries. In terms of agriculture, there were importers and exporters in both sides, making the respective positions of countries much more diversified. Thus, negations at the WTO, which in principle called for a unanimous decision among members, became stagnated. Even up until now the WTO has failed to deliver a conclusion.

A proposal that only a country like Japan can make

There were four main reasons why the grand goals of (1) coexistence of various forms of agriculture, (2) fulfillment of multiple functions of agriculture, and (3) securement of a stable food supply that Japan had proposed failed to bring countries together.

First, for exporters of agricultural produce, as long as agriculture developed, it would automatically lead to food security and fulfillment of multiple functions. Non-trade concerns were, from the beginning, a product of compromise born during the negotiations to promote free trade, and, as such, they did not attract exporters' interest much.

Second, fulfillment of multiple functions of agriculture only attracted countries that had no problem in food procurement. Developing countries cared less about it. Even if they were interested in it, arguments for such a non-essential goal would only shackle their negotiations.

Third, the specific non-trade concerns that each country was interested in were not necessarily uniform. For instance, the European Community showed a strong interest in farm animals' comfort (animal welfare) as a non-trade concern and submitted a proposal on animal welfare and agricultural trade. On the other hand, Japan, with many of its animal husbandry farmers rearing animals in narrow cattle barns and henhouses, chose to exclude animal welfare in its list of non-trade concerns in the negotiations.

The fourth reason was differences in perception of the relationship between agriculture and the environment. Sometimes called the greatest destruction of the environment by people, agriculture puts a tremendous burden on the environment. Take methane gas, the greenhouse effect of which is more than 20 times that of carbon dioxide, for instance. The largest emitter of methane is rice production. Is agriculture bad for the environment, then? The answer differs in accordance with the length of history of agriculture. In a country with only a short history of agriculture, such as the United States, the default condition is considered to be that without any human interference. In these countries, there is a strong belief that it is better for the environment to stop agriculture.

In countries with long histories of agriculture including Japan and western European countries, on the other hand, because their societies formed together with agriculture, the basic default condition to them, more often than not, includes agriculture. In these countries, many feel they are surrounded by nature away from urban bustle when they see mountain villages, forests studded with rice paddies, and reservoirs, even though these are maintained and sustained by human hands. This nature that has been kept and maintained only with the help of humans is quite different from such well-known wild landscapes as Yakushima Island, the Shirakami Sanchi mountains, or the Shiretoko peninsula.

The philosophy advocated by Japan could have been proposed only by a country like Japan, a high-income country with a long history of agriculture reliant on imports for food but which, nevertheless, can afford to pay attention to the non-food-related values of agriculture. Being such a unique existence, Japan found that attempts to find allies in the international negotiations did not go as smoothly as it had expected.

Value of rice paddies reassessed

Immediately before and after the third turning point in 1993, studies connecting Japanese identity with rice paddies began to appear. These included Sekiya Shin'ichiro's *Suiden no Hataraki* (*Workings of a Rice Paddy*) in 1992 (Sekiya, 1992), Tomiyama Kazuko's *Nihon no Kome* (Japan's rice) in 1993 (Tomiyama, 1993), and Ōnuki Emiko's *Kome no Jinruigaku* (*Anthropology of Rice*) in 1995 (Ōnuki, 1995).

In terms of policy, too, changes were urged. Even as the Japanese government suspended the policy to protect agriculture through price support in compliance with international trade rules, it needed to construct a new logic to stress the positive influence of rice production and rice paddies on the environment because it was now allowed to disburse subsidies for protecting the environment. While fields being in fallow was tolerated under the earlier policy to curb rice production, its contribution to the protection of rice paddies was re-evaluated in 1995. This led to increased subsidies for rice paddies fallowed under orderly control.

New rice production borne out of a sense of crisis

The third turning point of 1993, when the sense of crisis about rice imports intensified and people grew apprehensive about the future of rice production, was also a time when Japan's farmers began to pursue closer connection with rice consumers and urban dwellers; that is, the non-farming population.

Tsūjunkyō, one of the largest stone-built arch bridges in Japan, is located in Yamato-chō, Kumamoto prefecture, a mountainous and cool highland sandwiched by somma of Mt Aso and the Kyūshū Sanchi mountains. Built by village headman Futa Yasunosuke in 1854 toward the end of the Edo period, it is an aqueduct that provides water to the Shiraitodai plateau, where people had long suffered from famine caused by a dearth of water. Construction of the bridge led to cultivation of about 100 ha of new rice paddies.

One hundred and thirty years after the construction of Tsūjunkyō, locals of Yamato-chō faced a new crisis of the opening of the rice market. Believing that low uplands would be hit first by rice imports, they developed a regionwide new brand of rice called Tsūjunmai to help ensure the survival of rice production. At the same time, local farmers poured energy into pioneering closer relations with individual rice consumers. This kind of effort was still rare in those days. Yamato-chō's rice farmers offered consumers a set consisting of 1 shō (approx. 1.8 l) of rice for tasting, a pamphlet on Tsūjunmai, and a postcard to use to apply for a 1 year subscription to deliveries of the rice. These farmers aggressively sought customers with their own legs and enthusiasm.

Concerned for its rice terraces, where rice production is economically disadvantageous, Yusuhara-chō, Kōchi prefecture, in 1992 launched the first rice-terrace partnership program in Japan. Through this program, city dwellers who are interested in farm experiences and the preservation of rice terraces can rent rice paddies. They not only experience rice production, but also have a chance to mingle with nature and the local people. This partnership program utilizes values generated from rice production, aside from rice itself, and since 1993 has spread throughout Japan.

Resolution of the Rice Problem

The destination of produced rice

Before going into the details of changes in the legal system concerning rice, let us look into the situation of both the officially endorsed and unendorsed black market rice so prevalent in the market in those days. Officially endorsed black market rice, which had been first introduced in 1969, had increased to 4 million tons by 1990. Partly because the government had set the price at which it bought rice lower than its price for selling rice since 1987, a growing number of rice farmers found direct marketing of rice, including black market rice, more attractive.

Of all the rice produced in Japan in 1990, 42% became officially endorsed black market rice, and 23% became unofficially endorsed black market rice that rice farmers gave away or sold themselves. Together, these quantities occupied two-thirds of the entire rice harvest in that year (see Fig. 4.3), and amounted to 80% of all the rice distributed that year, excluding rice set aside by farmers for their own consumption and seed for the next season. And damage from 1993's cool summer triggered heightened consumer interest in purchasing rice directly from rice farmers.

Unendorsed black market rice being shipped out in Oogata-mura, Akita prefecture (*The Mainichi Shimbun*, February 8, 1990). (Used with permission from Mainichi Shimbun)

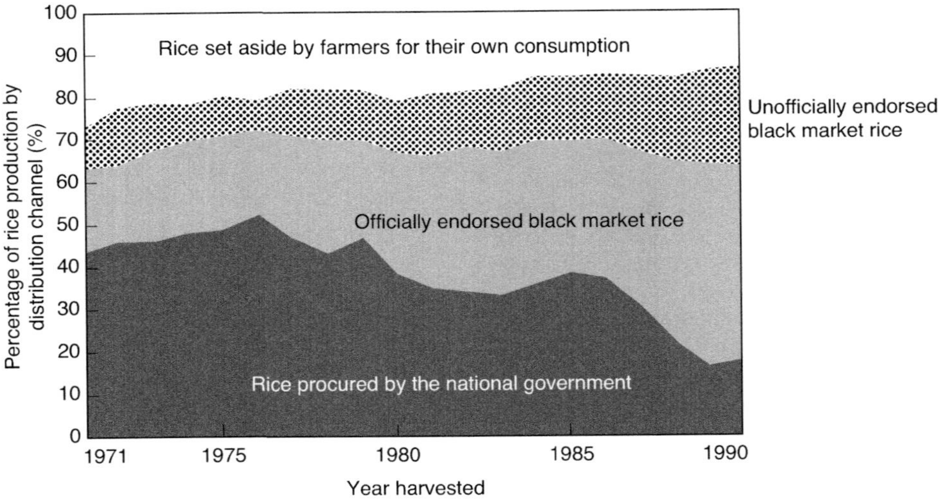

Fig. 4.3. Destinations of harvested rice. *Source*: based on the Food Agency's *Beika ni Kansuru Shiryō* [*Documents on Rice Price*]. *Note*: rice for processing is excluded.

Eclipse of rice control by state

The Staple Food Control Act of 1942, which had survived repeated amendments since the wartime, had finally yielded to the pressure of newly concluded international rules and, domestically, a loss of substance. While the government buying price for rice also had a function of supporting rice farmers' income, even that was lost when the purchase price became lower than the market price. The question, then, was when to abolish the act. Harihara Hisao, the then assistant director of policy planning at the Food Agency, reminisced as follows:

> Although nobody complies with the Staple Food Control Act and it has long lost its teeth, it is still believed that the Food Agency would reject its reform. Inside the agency, however, once they were liberated from the fulltime devotion to rice negotiations, many officials were determined to tackle reform of the act as a means to settle long-pending agency matters at one stroke. (Suga, 2020)

During the Uruguay Round, the Japanese government was fully committed to protecting Japanese rice at any cost. So, studying reform then was not easy; nor was it possible to assign officials for that purpose because they had to be devoted full time to the ongoing negotiations. Consequently, creating a new legal system was postponed until the conclusion of the international negotiations. The Staple Food Control Act, which was given a coup de grace at the conclusion of the Uruguay Round, was abolished in 1995.

Unregulated rice distribution

In place of the Staple Food Control Act, the Act on Stabilization of Supply, Demand and Prices of Staple Food (or Staple Food Act for short) was enacted in 1995. The act was reformed in large measure in 2004, resulting in the abolishment, in principle, of the government regulation of rice distribution that had occurred since pre-World War II days. Consequently, both rice farmers and rice distributors became free to sell and buy rice.

Even the concept of black market rice became passé now that rice farmers were free to sell rice directly to wholesalers, retailers, and consumers openly. A notification system, instead of a permission system, now applied to dealers in the business of shipping and selling rice. The time had come for everyone to sell rice as long as the seller met a few conditions. As mentioned earlier in this chapter, rice became a subject of tariffication in 1999. After that, anyone could import rice freely as long as they paid the tariff.

Now the state handled only stockpile rice and minimum access rice. The stockpile rice system was launched in response to the government's self-analysis on its inadequate actions during the very poor harvest in 1993. Today, the Japanese government stockpiles some 1 million tons so that it can cope with 2 consecutive years of once-in-a-decade levels of famine. The government imports some 770,000 tons of minimum access rice today.

Countermeasures for a flagging price of rice

In compliance with the new international rules agreed at the Uruguay Round, the Japanese government terminated its policy to support the price of rice. When as a consequence the price dropped, the government instituted a mechanism to pay subsidies directly to rice farmers to make up for the drop. The situation in Japan differed from that in the European Union, where introduction of the market mechanism and provision of subsidies to directly help producers were carried out at the same time as the export subsidy was abolished. In the European Union, the income standard applied for income support had already been raised beforehand with the export subsidy. In Japan, on the other hand, the rice price had already fallen by 1998 due to reduction of the government's rice buying price in 1987 as well as the presence of black market rice, which was influential enough to influence rice prices. There was, thus, a time lag between the drop in the price of rice and subsequent provision of the government subsidy.

For these reasons, the Japanese government's new mechanism failed to sufficiently make up for the rice farmers' reduced income. Moreover, because the amount of subsidy to be provided was calculated on the basis of the average over the previous few years, this meant that as the rice price continued to decline, so did the amount of subsidy.

Decreasing number of rice dealers

To consumers, a drop in the price of rice is a welcome change. Rice vendors, who used rice in eateries, boxed lunches, and onigiri rice balls, were particularly sensitive about the price of rice. Cheaper rice would create more opportunities for consumers to eat rice in eateries or in the form of ready-made meals in place of rice cooked at home.

Diversification of rice distribution also led to a drop in the rice price. Now consumers could enjoy cheap rice by purchasing it directly from rice farmers. As if to follow suit, supermarkets and consumers' cooperatives, too, purchased rice in large amounts and sold it cheaply through mass marketing. And taking advantage of their capital affiliates, mass retailers, the restaurant industry, and the home-meal replacement business started marketing rice in full swing. By going into the rice wholesale business, these entities came to exercise formidable influences on the rice distribution industry. These developments meant that traditional rice dealers were forced to relinquish their position as the main actors in rice marketing.

Earlier, under the Staple Food Control Act, the JA Group was categorized as delivery service. With the enactment of the Staple Food Act, however, the cooperative came to increasingly take on the character of a cooperative marketing business. Also earlier, the agricultural cooperative was responsible for storing rice on the prefectural level after collecting it from rice farmers. Now that this duty was lifted, it became possible for individual local chapters of the JA Group to market rice on their own or to sell it directly to wholesalers and retailers all over Japan. Also, exclusive rice in packaging that specified where the rice was grown, nowadays a common sight in supermarkets, was introduced to the mass market in about 1995. For instance, the Seiyu supermarket chain started selling Akitakomachi harvested in the Nishi-Senpoku region of Akita prefecture, while the Daimaru Peacock chain began to sell organic Koshihikari rice exclusively distributed by the Kurobe city agricultural cooperative of Toyama prefecture in its 30 branches in the Kantō region.

Encouraged by the Act on Stabilization of Supply, Demand and Prices of Staple Food, the number of farmers and supermarkets that started to sell rice spread rapidly. (*The Asahi Shimbun*, November 1, 1996). (Used with permission from Asahi Shimbun)

The rice problem resolved

These new legal systems resolved many of the problems concerning rice in Japan. As far as the day-to-day purchase and consumption of rice was concerned, people no longer felt inconvenience. In that sense, it would not be an overstatement to say that all the rice problems had been resolved. Besides, the Japanese government now stockpiled rice in order to cope with any future poor harvest. Japan became a society where people no longer worried about rice.

Today, a wide range of channels exist to obtain rice, from the grocery store to the home center to the many rice farmers, quickly found on the internet, who sell rice directly to consumers. Moreover, when consumer preferences shift to purchasing tasty rice or eating boxed lunches, onigiri rice balls, or in restaurants instead of cooking at home, rice farmers and distributors immediately react to produce better tasting rice, develop new breeds of rice suitable to these new uses, and pioneer new distribution channels. An environment has emerged today where it is easy for rice farmers with the will and capacity to respond to consumers' needs to be nurtured.

Of course, some consumers may demand cheaper rice and more tasty rice. Other people may criticize that distribution channels are not efficient enough or the tariff on overseas rice is too expensive. And it is problematic that a rice market that can set the standard for the price of rice has not yet been established. Still, these unsettled rice-related problems are all a matter of degrees. For instance, liberation of rice distribution provided a breeding ground for illegal reselling of minimum access rice and fraudulent claims of product origin. Subsequently, however, a new law was enacted that made it possible for buyers to specify the distribution route of their choice.

The basic framework of a society where rice is traded freely on the basis of the market mechanism was, thus, completed with the reform of legal systems triggered by the third turning point in 1993.

Grand Reform of the Policy to Curb Rice Production

Changing objectives of the policy to curb rice production

While the Staple Food Control Act was literally a law allowing the state to directly control rice, the new Staple Food Act was devoid of any reference to control. This signified the lifting of the role to directly control rice from the state. It should be noted, however, that the Staple Food Act took on the purposes of the Staple Food Control Act to maintain a balance between rice production and consumption and to stabilize the price of rice. And the policy to curb rice production was given legal basis under the new law. The Staple Food

Act stipulates that it is the state's role to carry out the policy to curb rice production, maintaining the balance between rice to be produced and rice to be consumed. It also stipulates that the subsidy for surplus rice is to be continued.

As discussed in Chapter 3, the policy to curb rice production began as a measure to lessen the financial burden on the state caused by rice surpluses when rice was directly controlled by the state. Therefore, once the Staple Food Control Act was abolished, the policy's aim became incongruent. Faced with this situation, the government stressed in its promotional activities that this policy was being carried out voluntarily by rice farmers to maintain the of rice price. This was nothing short of an attempt to replace the purpose of the policy.

It was the cooperative JA Group, an aggregator of massive amounts of rice, that was expected to play a particularly important role in this conjunction. Under the Staple Food Control Act, the government had been the main actor while the JA Group was assigned a secondary role. Under the Staple Food Act it became necessary to attempt a switchover of roles to make the JA Group the main actor to be supported by the government. The policy to curb rice production chiefly driven by the JA Group derailed within 2 years, however, because the size of the rice surplus grew unmanageable by the group.

Aside from consecutive good harvest years, the large rice surplus occurred for two reasons. First, in an imbalanced situation where free trade of rice became firmly established yet at the same time rice production continued to be discouraged, some rice farmers grew lax about curbing rice production. Second, when even a few among the many rice farmers overproduced rice, it would hardly affect prices. Thus, by producing more rice than assigned, some rice farmers were cunningly able to take advantage of other farmers who remained in compliance with the government instruction.

Also, now that it had become legitimate to openly deal with what used to be called black market rice, more and more rice farmers began to sell rice to purchasers other than the JA Group, weakening the cooperative's leadership over rice farmers. The government PR stunt to regulate rice production in order to maintain the price ended up losing its persuasiveness among rice farmers due to the drop in rice price caused by the rice surplus.

At this stage, restriction of rice production no longer benefited the country. It also ceased to function well for the benefit of rice farmers themselves. One simple question is: what is the purpose of discouraging the overproduction of rice in the first place? Or, what is the goal of the policy to curb rice production in the first place? In those days, even the Food Agency recognized that it had become difficult to comprehend who the policy to curb rice production was for and what it aimed to accomplish.

The way rice production should be in the first place

In order to make this policy to curb rice production easier to understand, a policy direction – the so-called rice policy reform that had been launched in full swing since 2004 based on deliberation at a private study group reporting to the director-general of the Food Agency – on the way rice production should be in the first place was set out. More concretely, the way rice production should be in the first place is explained as follows:

> To promote consumer- and market-oriented rice production by efficient and stable management units to ensure a stable supply of rice that fulfills various conditions, including price, required at each level of demand, starting with diversified consumer needs, and being mindful of producing sellable rice, astutely depicting trends in market demand.

The central pillar of this policy direction was development of managers who could astutely detect marketable rice. Promotion of marketable rice would also lead to prevention of the overproduction of rice.

Additionally, under the rice policy reform, the target year for which was 2010, the current state-led measure to curb rice production would be gradually shifted to a system driven chiefly by farmers and the JA Group. Also, in 2007, a measure to support the businesses of rice producers was introduced on the conditions that they took part in the policy to curb rice production and that their farmlands were larger than 10 ha in the case of Hokkaido, 4 ha in the case of other prefectures, and 20 ha in the case of collective farming. It was an attempt to single out large-scale farmers and help them to become economically self-reliant.

Rice paddy control based on a vision of ideal rice production

Rice production is a human act that combines rice and rice paddies. As such, attention should be paid not only to rice but also to rice paddies. However, while the policy direction on the way rice production should be in the first place describes in detail how rice distribution and rice farmers should be, its vision for rice paddies remains basically unchanged from the previous policy to curb rice production. The only difference was casual reference to fulfillment of multiple functions of agriculture and improvement of self-sufficiency. It can be conjectured that, through policy direction, the government intended to indirectly control rice paddy usage by upholding an ideal vision of rice, rice distribution, and rice farmers, instead of directly proposing an ideal vision for rice paddies.

Centering attention on rice instead of rice paddies significantly influenced the policy to curb rice production. The goal assigned to each rice farmer used to be the amount of rice paddy area where rice would not be produced. That now changed to the amount of rice to be produced. In addition, rice paddies not to be planted with rice used to be assigned to farmers through an administrative route from the national government all the way to prefectural and municipal offices. Now it is the JA Group, the entity in closest proximity to rice farmers, that determines the amount of rice to be produced; the government only provides relevant information that is helpful for each local chapter of the JA Group to conduct production adjustments on its own. After the new policy to curb rice production shifted attention from surplus rice paddy area to rice, the government's grasp of and involvement in the issue of dormant or abandoned rice paddies became weaker.

It should be noted that 'the way rice production should be in the first place' includes rice whole crop silage (WCS) as one kind of marketable rice. Rice WCS has been studied more, developed, and actually applied in certain areas in Japan since the second turning point in 1978. The price of rice WCS is so low that it cannot be produced unless government subsidy is provided. Inclusion of rice WCS in the government's list of marketable rice implies that the government may intervene in the definition of marketable rice itself; a foreshadowing of the subsequent arrival of the fourth turning point in 2008.

Vision for surplus rice paddies imposed on local municipalities

As for rice paddy usage, which was hardly touched on in the policy direction on the way rice production should be, a mechanism was created for rice farmers to receive government subsidies for curbing rice production. The subsidy program was to be carried out through municipal councils comprised of local government, farmers, and farmers' associations acting on their vision for rice paddy usage. Whereas previously uses for the government subsidy had been set uniformly throughout the nation as conditions for receiving the subsidy, the new policy allowed each municipality to determine on its own how it would use the government subsidy. This encouraged originality and ingenuity and for each region to come up with creative uses for surplus rice paddies. Not a small number of new local development programs have been tried under the new policy, including providing more discretionary subsidies to wheat, soybeans, and other crops into which the local community wished to pour energy, cultivating new buckwheat fields, and planting field mustard for better scenery and also rapeseed oil.

It should be noted, however, that this policy had an aspect of imposing on local communities the task of envisioning an overall picture for rice paddies, including surplus ones, which the national government could not deliver. The government's vision of what to do about surplus rice paddy area per se and rice paddies as a whole in the entire country – that is, its ideal vision for Japan's rice paddies – continued to remain obscure. Local chapters of the JA Group were instructed to function as the secretariat to the municipal councils. Letting local municipal councils determine how surplus rice paddies would be used and leaving disbursement of the government subsidy up to them confirmed the obscure situation about leadership in the policy to curb rice production; that is, whether it was the government or the JA Group.

The new policy also stipulated that the government subsidy to farmers who curb rice production would be continued at a fixed amount. The money was provided in the name of a subsidy, but in substance it was a measure to put a break on the government's financial burden in the face of a growing surplus of rice paddies. As such, the subsidy was not made to resolve the problem of surplus rice paddy area in the first place.

Although the era when rice and rice paddy issues had to be treated separately had arrived at the second turning point in 1978, it was assumed during the rice policy reform that the rice paddy problem could be solved when the rice problem was attended to. Yet due attention was not paid to the rice paddy problem even after the third turning point in 1993. While a policy was instituted in 2007 to encourage local communities to come together to preserve farms, farm roads, ditches, and reservoirs, the question of what to grow in rice paddies remained unsolved.

The Year 1993 in Retrospect

Once international rules on trade in agricultural produce were concluded, Japan's rice was partially liberalized, being protected only by the provision for non-trade concerns. The Staple Food Control Act was abolished in 1993; this liberated rice, in principle, from distribution control and other regulations, which resulted in freer trade of rice. Consequently, as far as buying and eating rice are concerned, the so-called rice problem for the Japanese people has been resolved on the fundamental level. What remains are the issues of brushing up individual rules and updating them in response to changing social conditions and restrictions.

At the second turning point, it became necessary to consider the rice problem and the rice paddy problem separately. In the policy carried out at the third turning point in 1978, however, the focus was clearly on rice instead of separating rice and rice paddy issues. It was hoped that, by liberating rice from distribution restrictions and supporting selected large-scale rice farmers, rice paddies would end up being better utilized. The rice policy reform, which begun in 2004 with 2010 as the target year, swung into full gear in 2007.

Meanwhile, have rice farmers been only pushed around by domestic and international policies? Not necessarily. Rice farmers have delivered their opinions on policy through the JA Group, by far the greatest pressure group in postwar Japan. Not only that, rice farmers have also expressed their choices through the ballot as sovereign citizens. In 2007, when the fully equipped policy commenced, the Liberal Democratic Party, which had relied on votes from rural areas as its power base, suffered a crushing defeat, which triggered the arrival of the next turning point in rice policy.

References

Ōnuki, E. (1995) *Kome no Jinruigaku* [*Anthropology of Rice*]. Iwanami Shoten, Tokyo.

Sekiya, S. (1992) *Suiden no Hataraki* [*Workings of a Rice paddy*]. Ie no Hikari Kyōkai, Tokyo.

Suga, M. (2020) *Heisei Nōsei no Shinjitsu: Kīman ga Kataru* [*Truth About Agricultural Policy in the Heisei Era According to Key Persons*]. Chikuma Shōbō, Tokyo.

Tomiyama, K. (1993) *Nihon no Kome* [*Japan's Rice*]. Chūō Kōronsha, Tokyo, Japan.

5 Emergence of the Concept of Full Utilization of Rice Paddies: The Fourth Turning Point in 2008

Abstract

Chapter 5 defines the year 2008 as the fourth turning point and expounds on changes in agricultural policy in Japan triggered by the 2007 House of Councillors election. In this election, the Liberal Democratic Party (LDP)/ Komeito coalition government lost a large number of seats to the Democratic Party of Japan, which had proposed a policy to increase the government subsidy for non-rice-producing rice paddies, resulting in a change of government in 2009. The chapter discloses that, even after the LDP/Komeito coalition returned to power later, the policy to expand the subsidy for such rice paddies continued. The chapter points out that even as the policy continued, it confronted limitations owing to budgetary constraints and the persistent decrease in rice-producing rice paddy area because of a dwindling population and the Japanese people's general disenchantment with rice.

Rice and Elections

Turning point of the rice paddy surplus problem

At the time of writing the Japanese version of this book in 2022, the ruling Liberal Democratic Party (LDP) held the reins of Japan's rice policy, having succeeded the rice policy that began in 2004.[1] The present rice policy is decisively different from that which existed in 2004–2007. It can be summarized as the difference in attention to the rice paddy surplus problem. The concept of full utilization of rice paddies emerged in 2008 and has been passed on as policy up to now. In retrospect, 2008 was a turning point.

Readers may think that the phrase *full utilization of rice paddies* means simply making full use of rice paddies. In actuality, however, things are a little more complicated. There is no single definition of the concept, and its meaning has been in constant flux since its first introduction. Before going into the issue of full utilization of rice paddies in this chapter, let us begin with the relationship between rice and elections.

Rice, subsidies, and political parties

After the end of the Pacific War, rice came to be closely associated with government subsidies and political parties. We can trace the origin of this to the severe treatment of rice farmers by Joseph Dodge, the financial adviser to the Supreme Commander for the Allied Powers, who was skeptical about the idea of subsidies. For instance, Dodge argued that rice farmers themselves should improve rice paddies and irrigation ditches on their own without the benefit of subsidies. Because rice paddies owned by any one rice farmer were often scattered throughout an area and irrigation ditches were communally owned, an individual farmer could do only so much to improve them. Therefore, rice famers grew highly discontented with Dodge's policy.

When the Korean War boom improved the Japanese government's financial situation, voices demanding protection of agriculture became louder, resulting in the increase of subsidies to farmers. What started with improvements to rice paddies and irrigation ditches grew to also include subsidies for the expansion of food production and structural improvements to agrarian villages. When the subsidy for surplus

DOI: 10.1079/9781800627093.0005

rice paddies was offered in compliance with the policy to curb rice production, government subsidy expenditures began to balloon at an unprecedented pace.

In Japan, where almost all farmers produced rice and the entire farming population shared the pain of surplus rice paddies, providing a subsidy for surplus rice paddies meant that the time had come for almost every farmer to receive some government subsidy. Additionally, other subsidies to agriculture and agrarian villages unrelated to rice also became an indirect support for rice production, because most of the recipient farmers did produce rice.

Tōhata Shirō, member of the Policy Board, Bank of Japan, remarked as follows in 1967, the year of the first turning point:

> There is no element of corruption whatsoever in the rice price decision. The higher the price is raised, the more farmers will profit. Its image is much cleaner than construction of a bridge or a road. Any Diet member must wish to contribute to it for a better image.[2]

Subsequently, pointing to the government's policy to refrain from raising its rice buying price, Watanabe Michio, minister of agriculture, forestry and fishery of the first Ōhira Masayoshi cabinet formed in 1978, the second turning point year, persuaded senior members of the ruling party that re-election would be difficult if a compensating subsidy was not offered to farmers. That way, he easily succeeded in creating a new ¥10 billion subsidy for agrarian, forestry, and fishery villages. Mutō Kabun, Watanabe's successor, also succeeded in securing a government subsidy for rural villages.

By raising the rice price – and, when that was not possible, by increasing subsidies to farmers – so as to win elections, the LDP succeeded in building a strong electoral power base in rural areas. Still, it has to be added that rural areas have not always been a staunch voting bloc, as exemplified by the 1989 House of Councillors election. One year after the United States/Japan agreement to liberalize beef and orange imports, LDP candidates faced a tough election even in conservative stronghold districts. One such candidate, Higaki Tokutarō, was defeated in Ehime prefecture where tangerines are a local specialty.

Crushing defeat of the LDP and surge of the Democratic Party of Japan

Let us advance to 2007. As explained in Chapter 4, that was the year LDP policy was changed to target large-scale farmers. Altogether 300,000 farmers, including those who participated in community farming, were targets for this policy.

In preparation for the House of Councillors election of July 2007, the opposition Democratic Party of Japan (DPJ) countered LDP policy by proposing a system of individual household income support for agriculture. It was a proposal to support all farmers who sold agricultural produce on the condition that they comply with the policy to curb rice production. The policy called for subsidizing farmers income from all produce, including rice, wheat soybeans, coarse cereal, and animal feed. It aimed to preserve all existing farmlands and help accomplish self-sufficiency in food by providing all-embracing support by the state to 1.72 million farmers, including rice farmers in deficit. In its manifesto, the DPJ estimated the budget for the proposed system to be ¥1 trillion, or three times more than the LDP's ¥360 billion.

Some rice farmers left out by the LDP's selective policy decided to give the DPJ a try and voted for the party in the July 2007 election. The LDP suffered a crushing defeat in rural single-seat constituencies as many votes there went to opposition parties. The DPJ, in contrast, achieved a historical, landslide victory.

Rice policy adrift

The DPJ's policy proposal was inadequate in terms of theoretical rationale as well as planning on how to finance the proposed programs. Nevertheless, the sweeping victory for the DPJ, which revealed the attractiveness of the policy for all-embracing support of farmers by the state, was shocking to the LDP, which had just begun its policy of supporting selected farmers. Besides, the rice price dropped sharply in 2007 mainly because of overproduction. It was a time when people became increasingly vocal about questioning the validity of the LDP's rice policy. In October 2007, only 3 months after its crushing defeat in the election, the LDP decided to conduct a thorough review of its rice policy. As a result, the revised LDP proposal was adopted as government policy at a highly unusual speed.

Consequently, it was decided that all farmers who collaborated with the policy to curb rice production would receive government subsidies. Also, the proposed reform to shift the main protagonist of the policy from the government to agricultural cooperatives was withdrawn to have the government remain as the chief promoter. In addition, the government cancelled its 2007 termination of the system to allot rice production goals to prefectural governments and revived the old system. Thus, the reform to select rice farmers and to shift the promoter of the policy to curb rice production from the government to farmers and farmers' associations became derailed in only half a year.

Surplus rice paddies become a voting bloc

Although rice policy reform had been pursued as a 7 year program since 2003, including the preparatory period, it fell through all of a sudden in 2007. It has become common to offer the rice-centered explanation for this experience that treatment of rice had changed under the influence of the short-term political situation. My view, however, is that, in fact, it was rice paddies that affected mobilization of the voting bloc and prompted a review of rice policy.

The system of individual household income support for agriculture proposed by the DPJ was attractive to rice farmers because it could make up for the loss of expected income from rice production. But what was much more appealing to rice farmers was that their income was ensured if they produced non-rice produce (including wheat, soybeans, coarse cereal, and animal feed) in surplus rice paddies.

Of all the rice paddies in Japan in those days, 30% were counted as surplus paddies, and rice production exceeded target amounts in 31 prefectures, or two-thirds of all prefectures in Japan. Many rice farmers were faced with the common worry of what to do with surplus rice paddies, regardless of the size of their operation. The standard for targeted selection in the LDP's proposal was not paddy area producing rice but the area of a farmer's entire farmland, including surplus rice paddies. The larger the scale of operation, the more surplus rice paddies a farmer possessed.

For their part, in order to receive as much subsidy as possible to produce wheat or soybeans, small-scale rice farmers had to fulfill the scale of operation standard by expanding on their own or by joining community farming. As a result, hastily formed community farming initiatives appeared all over Japan. Whether farmers expanded their scale of operation through self-expansion or by participation in community farming, the rice paddies thus acquired usually included not only those where rice could be produced but surplus rice paddies as well. This goes to show that, while the national government tried to advance its rice-centered reform of rice policy, all rice farmers, big or small, had common worries about what to do with surplus rice paddies.

Increased subsidies for surplus rice paddies were good news to organizations specializing in farming on rented surplus rice paddies, as well as to rice farmers commissioning necessary tasks to those organizations and to landlords renting paddies to rice farmers. But the widespread rumor that rice farmers who failed to meet the scale of operation standard would not receive any subsidy for surplus rice paddies aggravated the anxiety and confusion in agrarian villages.

Subsequently, subsidies for surplus rice paddies continued to increase as if surplus rice paddies were a voting bloc. During this process, a measure to depress the price of agricultural produce, such as the reform taken by the European Union, was not carried out. Subsidies for surplus rice paddies, which were meant to restrict over-production of rice – which in turn would contribute to the stability of the price of rice for meals – brought double merits to rice farmers in that they benefited from owning surplus rice paddies as well as rice paddies used to produce rice for meals. But from the viewpoint of consumers, subsidies imposed a double burden; that is, the tax to finance the program as well as the price of rice for meals, which would not go down.

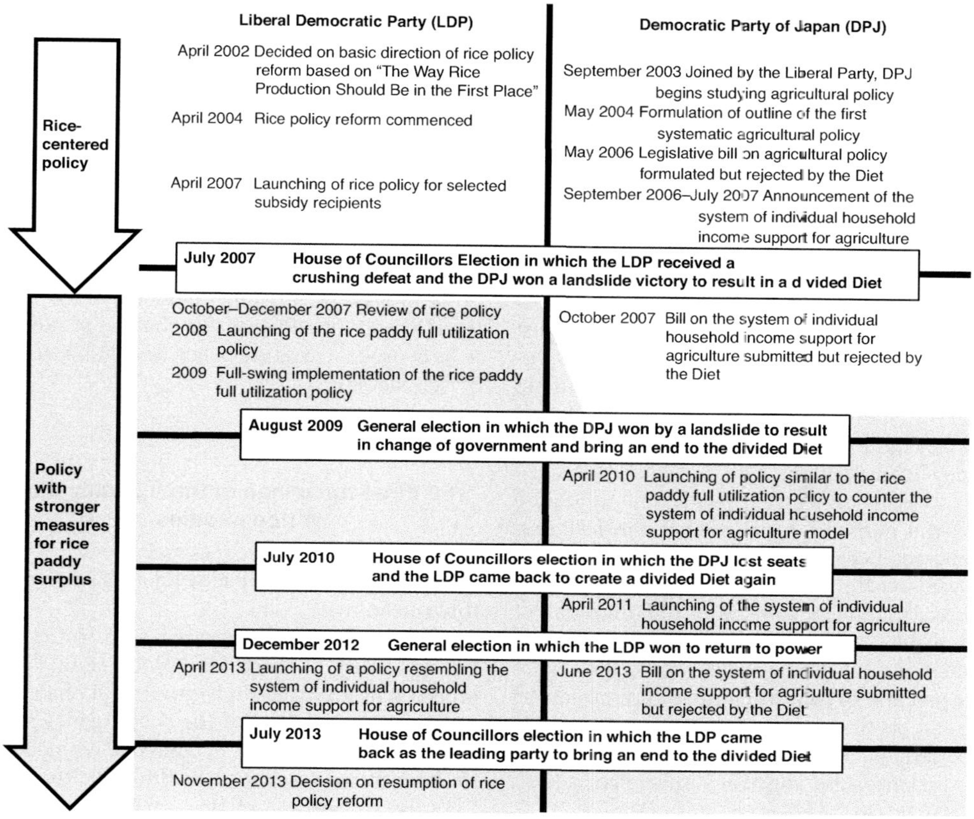

Fig. 5.1. Full utilization of rice paddy area as nurtured by the LDP and the DPJ (summary of events, 2002–2013).

Philosophy of Full Utilization of Rice Paddies

Full utilization of rice paddies promoted by the LDP and the DPJ

The idea for full utilization of rice paddies emerged from and was nurtured by the confrontation between the LDP and the DPJ. Let us review what happened (see Fig. 5.1). As we've seen, the LDP decided on the direction for its rice policy reform in 2002. The DPJ, which had urban voters as its power base, was joined in 2003 by the Liberal Party led by Ozawa Ichirō. After that, the party began to study a policy to prioritize rural areas and agriculture in hopes of taking over the government. The party's system of individual household income support for agriculture was drafted in 2006 when Ozawa became president of the DPJ. It was officially adopted in the party's manifesto in July 2007.

After winning convincingly in the July 2007 House of Councillors election, the DPJ rode the momentum to submit the bill on the system of individual household income support for agriculture in October. The rice policy in this bill, however, was heavily rice-centered – just like the LDP's policy in those days – and measures aimed at the rice paddy surplus problem were not fully discussed.

That same month of October the LDP promptly began reviewing its rice policy and decided to strengthen measures for the rice paddy surplus problem. As a counter to the DJP's system of individual household income support for agriculture, the LDP proposed raising compensation for surplus rice paddies in 2008 under the name of the measure for full utilization of rice paddies. That would be followed the next year by the full launching of the rice paddy full utilization policy, making 2009 the inaugural year of rice paddy full utilization.

Although the LDP beefed up its measure for the rice paddy surplus problem, its rice policy budget overall was smaller than that of the DPJ proposal. Meanwhile, momentum behind the surge in the DPJ's popularity did not stall; the party landed a crushing victory in the August 2009 general election to take the reins of government from the LDP in September. Interestingly, after the change of government, the DPJ adopted a policy quite similar to the LDP's policy for rice paddy full utilization.

The policy for rice paddy full utilization was the name of a policy that the LDP had proposed. Because of its substance, however, it was sometimes described that the DPJ had succeeded the LDP's policy for full utilization of rice paddies. Equally, the DPJ's own policy was sometimes called the policy for full utilization of rice paddies. In fact, when the DPJ reintroduced the bill on the system of individual household income support for agriculture in June 2013, it stressed measures aimed at surplus rice paddies, unlike the earlier bill proposed in October 2007. Thus, it can be said that the DPJ's rice policy emerged out of adopting the viewpoint on the full utilization of rice paddies that the party had inherited from the LDP.

When the LDP won a landslide victory in the 2012 general election and returned to power, it did not change the substance of its rice policy lest those who were directly involved should be confused, even though it did change the name. Consequently, in the July 2013 House of Councillors election, the LDP campaigned on the rice policy it had inherited from the DPJ. During the campaign, the LDP announced a policy to double farmers' income over the next 10 years. The LDP proposed to upgrade its measures to cope with the rice paddy surplus problem, referring to full utilization of rice paddies. Yet the LDP failed to show any concrete change of direction as far as policy toward rice itself was concerned.

When the LDP won the election and returned to being the no. 1 party in the House of Councillors after 6 years, ending the situation of a divided Diet, it announced new directions for its rice policy in quick succession. These reforms were tantamount to resumption of the rice policy reform that had been derailed in 2007. Unlike the pre-2007 rice-centered way of handling rice paddies, the LDP's new policy directions were characterized by the inclusion of enhanced measures to deal with the rice paddy surplus problem. Thus, the common perception of the need to tackle the rice paddy surplus problem was nurtured by interactions between the LDP and the DPJ over two changes of power to be consolidated as the policy of full utilization of rice paddies.

The three meanings of the full utilization of rice paddies

Roughly speaking, rice paddy full utilization has three meanings.

The phrase first appeared as a synonym for the policy to curb rice production. First proposed by a private study group that reported to the director-general of the Food Agency, the idea was later adopted by the LDP as policy. Its essence was a change in thinking; that is, from the suppression of rice production by the government to the prevention of overproduction of rice at farmers' own judgment. Farmers were encouraged to use their judgment in producing rice, after being enlightened about the attractiveness of agricultural produce other than rice for meals. Full utilization of rice paddies sounds much better, even positive, compared to area reduction or production adjustment of rice. The phrase was proposed as a slogan to symbolize the shift to new thinking.

Second, the phrase is an appeal to make as wide use of rice paddies as possible. Behind this appeal was the smaller-than-expected harvest of wheat, soybeans, and animal feed planted in surplus rice paddies, which will be discussed in detail in the following section.

Third, the phrase was a proposal to plant as much crop as possible in the same rice paddy so as to utilize the full potential of each paddy. The DPJ instituted a new government subsidy in 2010 to promote double cropping; the LDP continued the policy after it returned to power until 2016.

As can be seen, the relative weight of each of these three meanings of the full utilization of rice paddies has shifted along with the passage of time until today.

From subsidies for rice to subsidies for rice paddies

With its policy of full utilization of rice paddies, the LDP aimed at expanding farmers' choices in the use of surplus rice paddies by upgrading subsidies for rice for animal feed in addition to subsidies for wheat and soybeans. The intention was to mitigate farmers' sense of the burden imposed on them by the policy to curb rice production through the offer of an additional 'carrot' for surplus rice paddies.

In contrast, the DPJ's system of individual household income support for agriculture intended to eliminate the 'stick' that remained in the policy to curb rice production. In addition, the DPJ aimed to dissuade rice farmers from producing rice by providing a subsidy of ¥150,000 per ha exclusively to rice farmers who complied with the policy to curb rice production, in proportion to the rice paddy area in which rice for meals was planted, on top of the subsidy for surplus rice paddies. The DPJ's strategy was to explicitly stress that the policy to curb rice production was by no means coercive. At the same time, the DPJ wanted to mitigate farmers' sense of burden by increasing the amount of subsidy.

It has to be noted here that carrot and stick are two sides of the same coin and, as such, a carrot, too, can be coercive, as explained in Chapter 3. The higher the subsidy for surplus rice paddies, the less acute the sense of burden became on the part of rice farmers. But at the same time, the state's involvement in rice production became stronger. Although both the LDP and the DPJ proclaimed that their measures would weaken the air of coercion surrounding the policy to curb rice production, the perception still persisted.

The subsidy budget for surplus rice paddies began to increase when the subsidy for fiscal year 2008 was provided under the fiscal year 2007 supplemental budgets (see Fig. 5.2). The DPJ, with its policy launched in fiscal year 2010, carried out (1) abolishment of the stick, (2) addition of a new carrot in the form of a subsidy for the production of rice for meals, and (3) continuation of the carrot in the form of subsidies for surplus rice paddies.

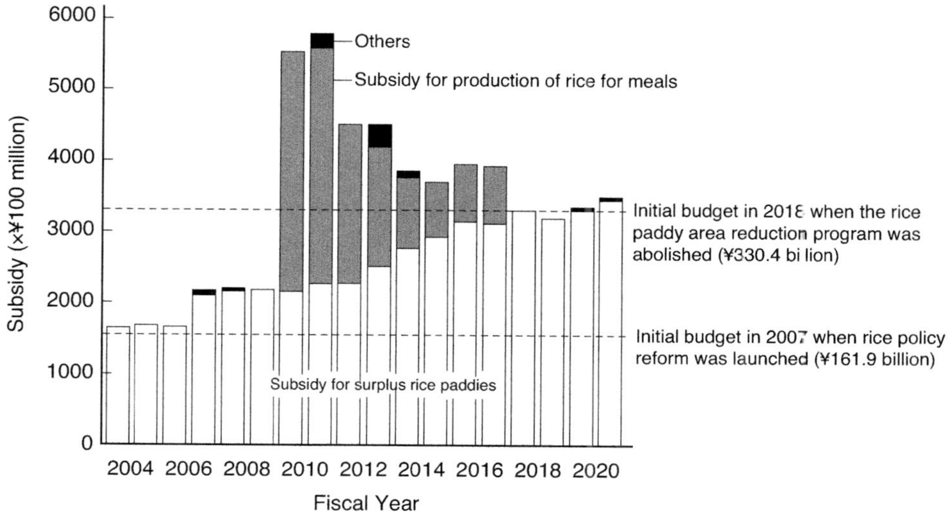

Fig. 5.2. Changes in subsidy to curb rice production. *Source*: compiled by adding the Ministry of Agriculture, Forestry and Fisheries' initial and supplementary budgets formulated in each same fiscal year. It should be noted that some of the supplementary budgets might be applied to production in the ensuing year.

When it returned to power, the LDP continued abolishment of the stick but also gradually removed the carrot. The subsidy for the production of rice for meals was reduced in stages until its eventual abolishment in 2018. As a result, after 2018, the only policy measure remaining to curb rice production was subsidies for surplus rice paddies.

The LDP government resumed rice policy reform in 2014 and announced abolishment of the rice paddy reduction program in 2018. This was followed by abolishment of the practice to allot rice harvest targets to prefectures. Reform at this time was the same in content as the reform carried out in 2007. What differentiated the derailed 2007 reform and the successful 2018 reform was the size of the subsidy for surplus rice paddies. The 2018 reform provided twice as much subsidy as the 2007 reform. Although a reform called abolishment of rice paddy reduction was carried out in 2018, the policy to curb rice production still remained strongly under government control.

Today's financial support for rice and rice paddies

At this point, let us confirm the latest scale of financial support for rice (see Table 5.1). First of all, there is no subsidy specializing in rice production. The only financial supports specifically for rice are aimed at the minimum access rice, in which the government is directly involved, and the management and trading of stockpile rice. For these, altogether ¥80 billion was disbursed. As for the subsidy specifically for rice paddies, ¥300 billion was provided to promote full utilization of rice paddies.

Aside from the financial support exclusively for rice and rice paddies, other government subsidies include compensation for income loss due to depression of rice and wheat prices, as well as subsidies for rice paddy and field consolidation. It is difficult to single out financial supports exclusively for rice and rice paddies, but it can be said that, overall, the total subsidy for rice paddies is larger than that for rice per se. Of

Table 5.1. Types of financial support for rice and rice paddies and their amounts. *Source*: based on documents prepared by the Fiscal Reform Subcommittee, Fiscal System Council. *Note*: figures are approximate amounts of the average annual disbursements immediately before and after 2021.

		Exclusively for rice and rice paddies	Other non-exclusive subsidies	Total
Rice		Management and marketing of minimum access rice (¥35 billion), Management and marketing of government stockpile rice (¥45 billion)	Income compensation when the rice price falls (¥10 billion)	¥90 billion
Rice paddies	Where rice is actually grown	None	Improvement and renewal of farmland and irrigation facilities (¥600 billion), joint activities to support multi-functional roles of farmland (¥50 billion), maintenance of agriculture in hilly and mountainous areas (¥25 billion)	¥975 billion
	Surplus rice paddies	Subsidy for rice paddy full utilization (¥300 billion)		
Total		¥380 billion	¥685 billion	¥1065 billion

those subsidies aimed at rice paddies, the large amount of subsidy for full utilization of rice paddies is particularly conspicuous.

Food crisis in 2007–2008

A backdrop to the increase of tax money spent on surplus rice paddies in the 2007–2008 period was, among other factors, the worldwide food crisis. The international price of cereals rose sharply during those years. The price of corn in particular gravely impacted Japan's livestock farmers, who relied heavily on imported animal feed. Coupled with the influence of the increase of corn's use as a source of bioethanol for major producers including the United States and Brazil, the situation triggered the full-fledged provision of subsidies for rice for animal feed and for rice flour to be processed into bread and noodles.

For Japan, having experienced the food crisis in the 1970s, maintenance of surplus rice paddies had an aspect of enhancing food security. The need to enhance food security came to be perceived more acutely upon enactment of the food security-conscious Basic Act on Food, Agriculture and Rural Areas and the arrival of a new food crisis.

Although criticism from the United States, a rival producer of animal feed in the form of corn, had been expected when the Japanese government decided to support rice for animal feed with government subsidies, such criticism has not surfaced to date. This may be because of the increase in the need for corn for bioethanol in the United States and the rise of an animal husbandry industry in countries other than Japan that relied on imported corn. While corn produced in the United States had been used mostly to feed to animals, in around 2011 – the time of the fourth turning point (2008) in Japan – U.S.-grown corn came to be used more for industrial purposes, including ethanol, than as animal feed.

Role of domestically produced animal feed in the human environment

Domestic production of animal feed not only played a role in the enhancement of food security for Japan, it also contributed to reducing the environmental burden. Animal dung and urine is the no. 2 industrial waste in Japan after polluted sludge. Three laws were enacted in 1999 to reduce the impact of animal dung and urine on the environment. That same year also saw the start of the government subsidy for production of rice whole crop silage (RWCS) and pasture using farmyard manure. With these developments, voices calling for the promotion of domestic production of animal feed grew louder around the fourth turning point.

Let us consider nitrogen's impact on the environment as an example. For agricultural produce, nitrogen is a nutrient. But nitrogen unabsorbed by produce remains in the soil to be drained into groundwater and rivers in the form of nitrate compounds. Water with a high content of nitrates can cause methemoglobinemia (also known as blue baby syndrome) suffocation in infants and can lead to the eutrophication of lakes and inland bays. In this way, nitrogen can have an adverse effect on the human body as well as the environment.

Among the countries in the world, Japan imports a phenomenal amount of nitrogen through food and animal feed. Because the animal husbandry industry imports a massive amount of animal feed and agriculture uses a huge amount of chemical fertilizer, Japan ends up holding a tremendous amount of unutilized nitrogen that becomes a cause of environmental contamination, the impact of which on groundwater and rivers through animal dung and urine is grave. For this reason, producing animal feed domestically puts less burden on the environment than imported animal feed.

New-Demand Rice: An Adrenaline Shot for Utilizing Surplus Rice Paddies

Stagnated wheat and soybean production

There were four takeaways from the meeting of the private study group of the director-general of the Food Agency, from which the term *full utilization of rice paddies* was born.

First, the group recognized the importance of making full use of surplus rice paddies in

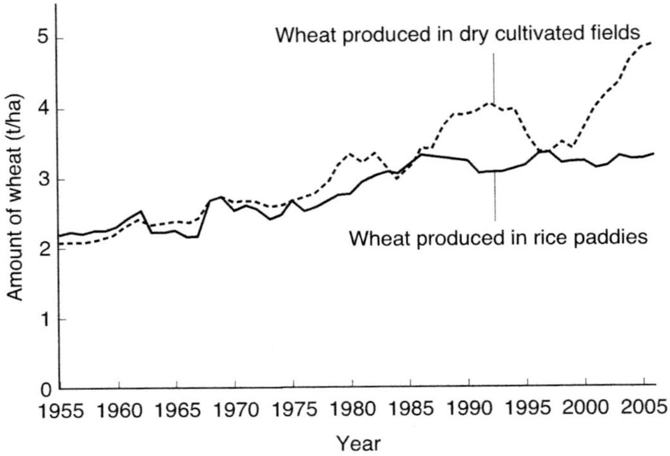

Fig. 5.3. Changes in crop amounts for wheat per 1 ha. *Source*: compiled based on 5 year changes obtained from the Ministry of Agriculture, Forestry, and Fisheries' *Sakumotsu Tōkei* [*Statistical Survey on Crops*].

order to provide a continuous supply of food to the Japanese people into the future. While the general inclination in those days was to believe that the subsidy for surplus rice paddies was for the benefit of rice farmers, in fact the private study group deemed the subsidy, which was an effective measure to curb rice production, as necessary for Japan as a whole.

Second, study group members shared the perception that some regions and rice paddies were not suitable for wheat and soybean production. While it is difficult to single out specifically which regions or rice paddies are or are not suitable for wheat and soybean production, because upland field crops such as wheat and soybeans are vulnerable to moisture damage, it might be challenging, generally speaking, to grow them in rice paddies with relatively poor drainage, particularly with climatic changes causing torrential rainfalls.

The per-area harvest for wheat tends to be smaller in rice paddies than in cultivated dry fields (see Fig. 5.3). Admittedly, some farmers planted wheat in their surplus paddies just to receive the government subsidy and didn't pay due attention to the crop. Still, the fact remains that the gap in harvest yield has certainly widened. As for soybean planting, it has been actively pursued in prefectures where parcellation and water drainage for rice paddies have advanced, starting with Saga prefecture (see Fig. 5.4). Conversely, soybean production tends to be inactive in prefectures where parcellation and water drainage for rice paddies have slackened.

Third, the study group encouraged filling rice paddies with water so as to ensure full utilization of rice paddies. In those days, the majority perception of the LDP's Subcommittee on Basic Agricultural Policy was that it was more important to use rice paddies as rice paddies than for anything else. But producing rice not for meals in surplus rice paddies has a few merits. For instance, the machinery and technology used to produce rice for meals can be used as is. Also, rice production causes almost no monocropping difficulty, which allows for continuous use of rice paddies. At the study group, these merits were emphasized.

Fourth, the study group proposed that rice which could not be produced without government subsidy, including RWCS and rice for animal feed, also should be included in promotions as marketable rice. This recommendation followed the line of argument outlined under The way rice production should be in the first place, in Chapter 4.

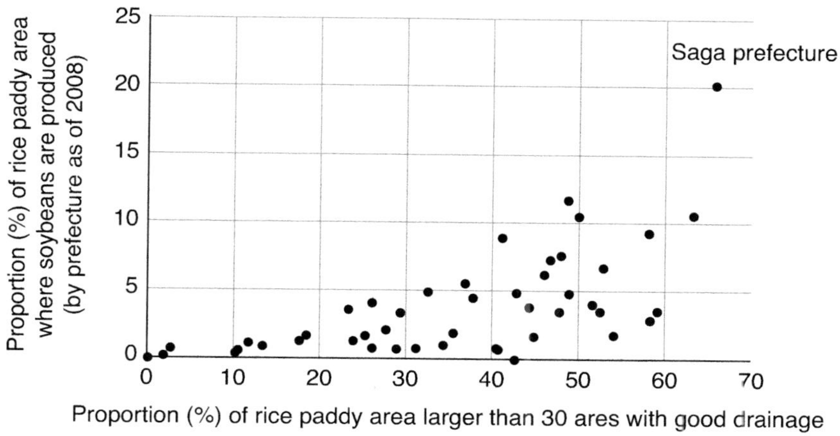

Fig. 5.4. Status of rice paddy development and soybean production. *Source*: based on Fig. 3.70 in the Ministry of Agriculture, Forestry and Fisheries' *2009-nendo Shokuryō, Nōgyō, Nōson Hakusho* [*Annual Report on Food, Agriculture and Rural Areas in Japan in Fiscal 2009*].

Rice processed into bread and animal feed

Just as the production of rice other than rice for meals had stagnated, so did the production of rice for processing, which had been introduced around the second turning point year (1978). In order to cope with the situation, rice for animal feed, RWCS, and rice for rice flour began to be promoted in 2008 under the name 'new-demand rice.' As a result, production of these kinds of rice rose (see Fig. 5.5).

In Fig. 5.6, the price of rice for bread is compared with that of imported wheat, and the price of rice for animal feed is compared with that of imported corn. A rice farmer spends ¥160,167 per ton to produce rice for bread or rice for animal feed, but bread is sold only at ¥80,000 per ton and animal feed at ¥40,000 per ton (see Fig. 5.6). The government subsidy, which makes up for the gap, has made possible the use of rice in ways that would be unthinkable under normal circumstances.

Rice for animal feed survived two changes of government

Almost all the income a rice farmer receives is from government subsidies (see Fig. 5.7). Particularly typical among new-demand rice is rice for animal feed. It would be only natural for the use of rice as animal feed, which cannot exist without government subsidy, to be the first to be criticized as a waste of tax money. But it has nevertheless survived the turbulence of two changes of government.

The government disbursed ¥500,000/ha in 2008 with certain conditions attached; this was raised to ¥550,000 across the board for farmers producing rice for animal feed throughout Japan in 2009. The amount was further raised to ¥800,000 in 2010–2013. Since 2014, the amount has been set between ¥550,000, at the minimum, and ¥1.05 million, at the maximum. Along with other additional subsidies, the government in total poured ¥77.2 billion into subsidies for 80,000 ha of rice paddies in 2018. This was a quarter of the ¥298.6 billion disbursed to curb rice production.

In 2009, a new law concerning rice for animal feed was enacted. In contrast to the slow expansion of surplus rice paddy area planted with wheat, soybeans, and pasture, the area for rice for animal feed began to increase rapidly in the fourth turning point year (2008). Together with RWCS, the area for rice for animal feed planted in rice paddies shot up in one bound to be almost equal with that for soybeans and wheat (see Fig. 5.8).

Books on this subject published around the fourth turning point include *Shiryōyō Mai no Saibai/Riyō* (*Harvest and Use of Rice for Animal Feed*) (Ozawa and Yoshida, 2009), co-authored

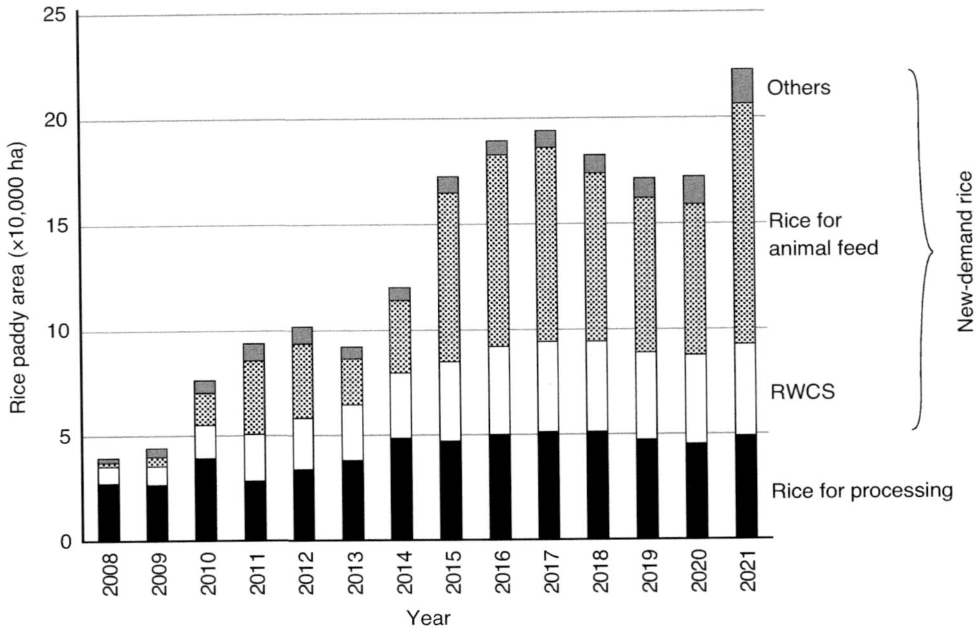

Fig. 5.5. Changes in rice paddy area where rice for processing and new-demand rice are grown. *Source*: based on the Ministry of Agriculture, Forestry and Fisheries' *Shinki Juyōmaito no Yōtobetsu Sakuzuke/Seisan Jōkyō no Suii* [*Changes in Planting and Production of Rice, Including the New-demand Rice, by Use*]. RWCS, rice whole crop silage.

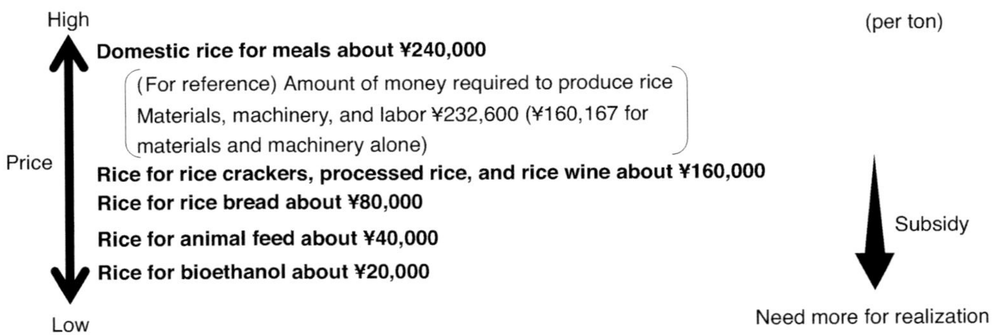

Fig. 5.6. Price and need for rice subsidy depending on usage, 2008. *Source*: based on documents submitted to a study group on rice systems centered around marketing as well as the Ministry of Agriculture, Forestry and Fisheries' *Nōgyō Keiei Tōkei Chōsa* [*Statistical Survey on Farm Management and Economy*].

by agricultural economists Ozawa Wataru and Yoshida Norio, who have long studied RWCS and rice for animal feed on the basis of cases in Yamagata prefecture. Other books by agricultural economists are *Shiryōmai no Seisan to Buta Nikushitsu no Kōjō* (*Production of Rice for Animal Feed and Improvement of Pork Meat Quality*) (Kumagai and Ohtani, 2009) and *Suiden Katsuyō Shinjidai* (*New Age of Rice Paddy Use*) (Taniguchi *et al.*, 2010).

The production target of rice for animal feed was first set out in 2010, in the basic plan

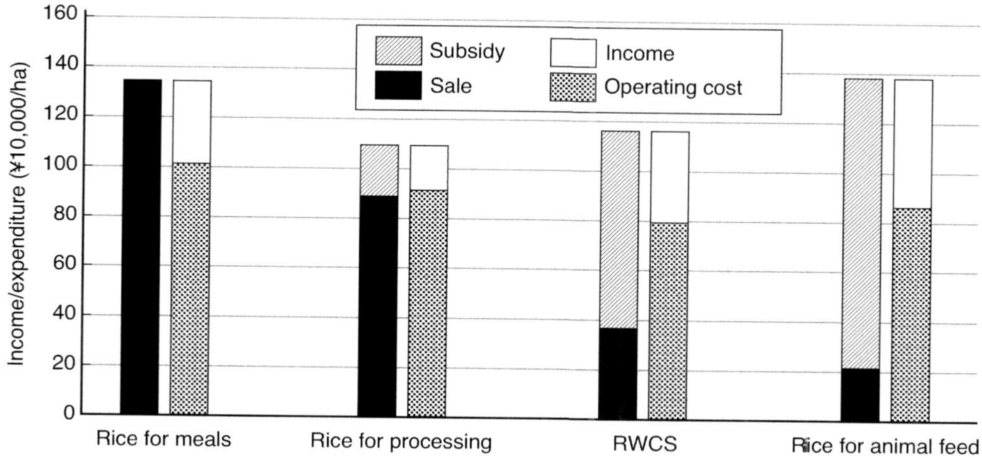

Fig. 5.7. Income and expenditure of rice for meals, rice for processing, RWCS, and rice for animal feed. *Source*: based on the Ministry of Agriculture, Forestry and Fisheries' *2021-nendo Keiei Shotoku Antei Taisaku-tō no Gaiyō* [*Fiscal Year 2021 Outline of Measures to Stabilize Business Income*].

based on the Basic Act on Food, Agriculture and Rural Areas. In 2020, the target for production in 2030 was set at 700,000 tons.

Specialized breeds of RWCS and rice for animal feed have been developed, but rice for meals has also been used for this purpose. The increase of rice paddy area used to produce RWCS and rice for animal feed since 2008 was not because of the introduction of new breeds. It can be exclusively attributed to the increase of government subsidies.

Characteristics of rice for animal feed

Different parts of the rice plant are harvested for RWCS and animal feed. In the case of rice for animal feed, the berry portion of the rice plant is used to feed cows, pigs, and chickens. In the case of RWCS, the stalk, leaf, and premature berry of the rice plant are all harvested to be fermented before being fed to cows (see Fig. 5.9). More rice for animal feed than RWCS is produced because the government subsidy for rice for animal feed is relatively high and rice for animal feed is easier to process, as outlined below.

To prepare RWCS, generally speaking, a farmer needs to harvest the whole rice plant in its entirety, roll it up, and wrap it up. This requires a machine specifically designed for this task. It also calls for certain skill, without which RWCS will get moldy and become useless. Moreover, RWCS is full of water when harvested, making one roll weigh more than 200 kg. This makes RWCS difficult to transport over long distances. It also calls for temporary storage spaces for efficient transportation. Because the quality and weight of RWCS vary from roll to roll, its distribution is basically confined to short distances.

In contrast, rice for animal feed can be harvested in the same way as ordinary rice for meals. Since it is granulated, it can be easily divided into smaller portions to make it easy to transport. It will be even lighter when dried. Various methods of processing rice for animal feed, including what farmers need to do at the time of harvesting, have been studied. Some methods are pound-down grinding, breaking down, heating, and fermenting. In actuality, however, as much as 76% of rice for animal feed is sent to processing plants, liberating rice farmers from the need for special machines or skill. Thus, rice for animal feed is much easier to handle.

I hasten to add, however, that easy to handle does not necessarily mean easy to adopt. Special machinery is required to harvest RWCS and prepare it for animal feed. Having invested in procuring machinery, farmers find it difficult to quit growing RWCS right away. Also, because

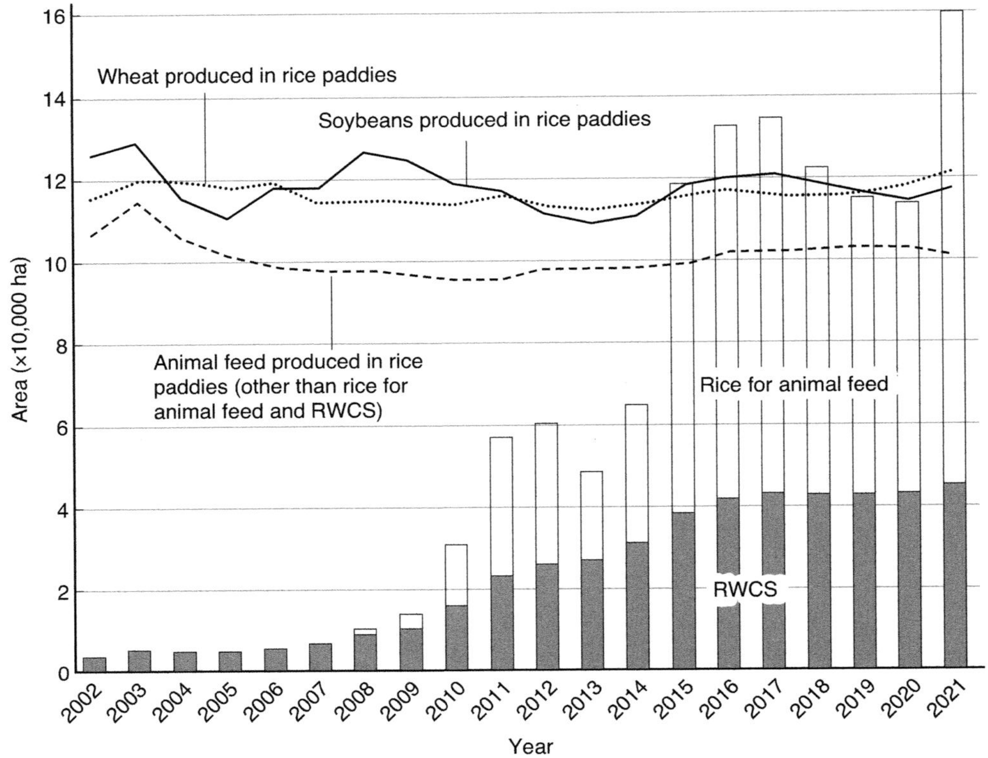

Fig. 5.8. Rice paddy area producing wheat, soybeans, and animal feed. *Source*: based on the Ministry of Agriculture, Forestry, and Fisheries' *Sakumotsu Tōkei* [*Statistical Survey on Crops*] and *Shinki Juyōmai-tō no Yōtobetsu Sakuzuke/Seisan Jōkyō no Suii* [*Changes in Planting and Production of Rice, Including the New-demand Rice, by Use*].

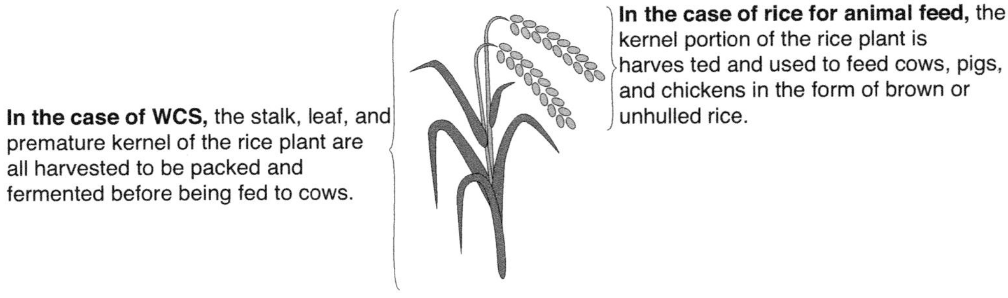

Fig. 5.9. Differences between rice for animal feed and whole crop silage (WCS).

the quality of RWCS and its distribution system become stabilized over time with experience, it becomes easier to continue the operation, once started.

On the other hand, obstacles for rice farmers to start or quit producing rice for animal feed are relatively small. For this reason, some rice farmers grow rice for animal feed only during periods when prices of other produce, including rice for meals, are low. Although some rice farmers continue to grow rice for animal feed because of the government subsidy, recent years have seen quite significant year-to-year fluctuations in its production.

(a) Rice for WCS is harvested in its entirety and rolled up. (b) Harvested rice is wrapped to produce rice WCS. (Photographer: Masayuki Ogawa)

A variety of processing methods have been tried for rice for animal feed. *Source*: National Agriculture and Food Research Organization (ed.) *Shiryō-yō Mai no Seisan/Kyūyo Gijutsu Manual 2016-nendo ban [Manual of Production and Supply Technologies for Rice for Animal Feed,* 2016 edition]. Photos are provided by Yasunari Yamamoto, Mie Central Agricultural Improvement and Extension Center, and Makoto Seki, Dairy Technology Research Institute of the National Federation of Dairy Cooperative Associations. (Used with permission)

The true problem of rice for animal feed

It is sometimes said that the high government subsidy for rice for animal feed is problematic. As long as a subsidy is well balanced against the social merits that rice for animal feed brings with it, however, the subsidy being high is not itself a problem. The fundamental problem of rice for animal feed, instead, is the obscure role of its production. In other words, for what purpose is rice for animal feed being produced?

The subsidy for growing rice for animal feed is set at a level that is not dissatisfactory in comparison with the income a farmer can expect when growing rice for meals. The amount of subsidy is by no means based on a calculation of the social role that rice for animal feed can play. On the other hand, the role of rice for animal feed has not been clarified to date. Yet it must be pointed out that rice for animal feed has come to play increasingly more roles as the years go by.

Rice for animal feed was recognized as an alternative for crop conversion by the Rice for Animal Feed movement in 1981, and a government subsidy of ¥200,000 per ha was newly instituted in 1987. Because the subsidy was too small, however, rice for animal feed was produced only in 500 ha at most in the 1980s and 1990s; an almost non-existent amount considering the entire area of Japan.

The situation changed drastically in 2004. When it became possible to set the government subsidy for surplus rice paddies in each region independently, full-fledged production of rice for animal feed began to be promoted in some regions. The rice for animal feed project started in Yuzamachi, Yamagata prefecture, after the area was certified as a food self-sufficiency special district, spearheading the nationwide spread of rice for animal feed. Rice for animal feed, which was an unnatural produce, so to speak, made possible only with government subsidy, subsequently spread to other areas as more reasons for government subsidies were invented. Over time, the rice paddy area planted with this crop also expanded.

Around the fourth turning point (2008), rice for animal feed was advocated as a crop that could utilize surplus rice paddies that could only produce rice. The LDP's earlier policy on full utilization of rice paddies advocated rice for animal feed paired with measures to deal with completely unused rice paddies and abandoned cultivated land.

Because the price of corn remained high due to the food crisis, the subsidy for rice for animal feed gained an additional role: rescuing destitute livestock farmers. Also, appeals to recognize other merits of rice for animal feed from the viewpoint of food security were made. For instance, growing rice for animal feed does not do any damage to surplus rice paddies, and rice for animal feed can be eaten by people if need be. Rice for animal feed as a means to pull up rice farmers' income was also advocated in the LDP's income-doubling plan for farmers announced during the July 2013 House of Councillors election campaign as well as in the domestic groundwork done in preparation for the Trans-Pacific Partnership (TPP). In 2014, the government subsidy for rice for animal feed was upgraded as part of the review of the government's rice policy reform.

However, as of 2022, the policy to curb rice production also advocates a shift from growing rice for animal feed to growing rice for processing, rice for export, and growing wheat or soybeans. This was a result of emphasis put on the role to reduce production of rice for domestic meal consumption among the various roles of the policy to curb rice production. The significance of domestic production of animal feed, which had been stressed after the food crisis in 2007–2008, is underemphasized here.

As we've seen, the role of rice for animal feed has not been clarified, nor has the interconnection among the various possible roles for rice for animal feed been resolved. To further confuse the issue, any role can be added or eliminated in accordance with the tendency of the times.

The gravest problem of the subsidy for rice for animal feed is not the heavy financial burden it brings but the fluidity of its reason of existence. In other words, the fact that the reason to provide a subsidy for rice for animal feed in the first place fluctuates with the tendency of the times is the problem. This unique feature of the subsidy for rice for animal feed, however, is not a disadvantage. On the contrary, the fluidity of the subsidy's reason for existence allows unrestricted proliferation of reasons to maintain the subsidy, which in turn provides a major justification of the subsidy for rice for animal feed (see Fig. 5.10).

The different roles of rice for animal feed for different farmers

Today, basically a uniform subsidy is set across the board for production of rice for animal feed. Because the reason for growing rice for animal feed is not clearly established, however, individual farmers throughout Japan find their own reasons for growing rice for animal feed, leading to unique problems in each region.

The wet paddies abundant in Daitōchō (present-day Ichinoseki city) of Iwate prefecture are not suitable for upland field crops, including wheat and soybeans. Therefore, under the policy to curb rice production, unplanted rice paddies multiplied. To cope with this situation, the Daitōchō city government studied rice for animal feed. In 2003, Daitōchō experimented with growing rice for animal feed in a 0.1 ha paddy. By 2007, the rice paddy area used to grow rice for animal feed had risen to 10 ha. Following the rise in the government subsidy in 2008 that area rapidly grew to 20 ha in 2008 and 30 ha in 2009.

Introduction of the new option to use rice paddies to grow rice for animal feed had a tremendous impact on community farming operators, which had been collecting rice paddies. According to a 2013 survey on the status of community farming, more than 25% of respondents said they grew rice other than for meals. Rice other than rice for meals was named the no. 1 earning produce in 3.8% of collective farming operations. While the number of collective farming groups planning to grow new crops decreased, those planning to grow new-demand rice for the first time, including rice for animal feed, greatly outnumbered those choosing to grow other produce.

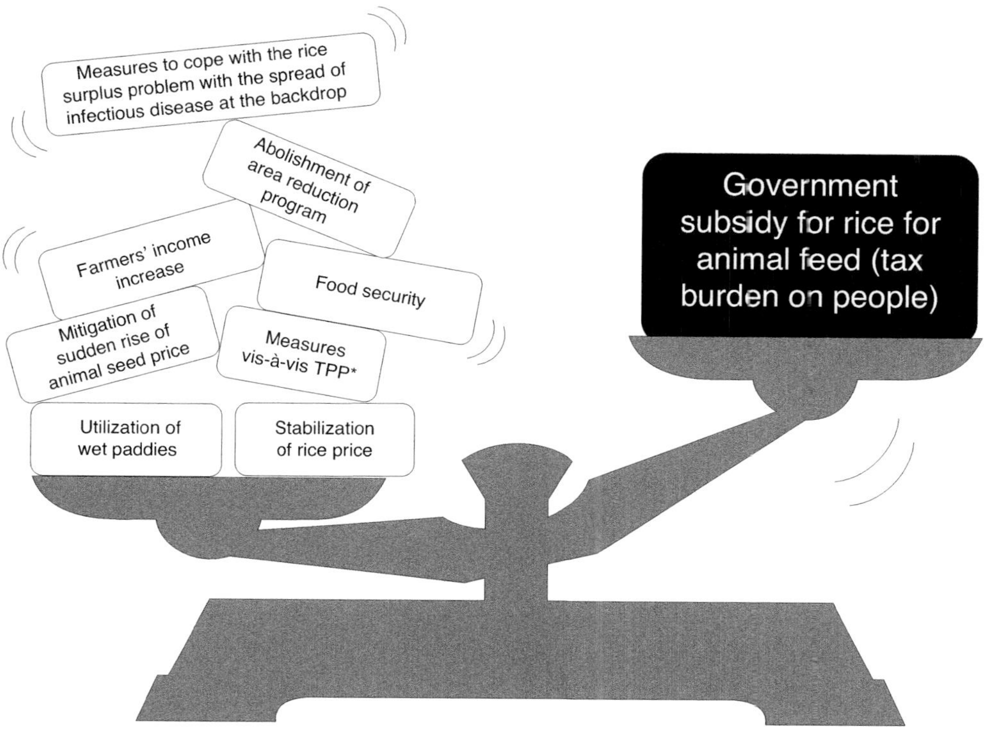

Fig. 5.10. The roles of rice for animal feed and government subsidy. *Note*: while rice for animal feed plays increasingly larger roles, the priorities and relations among them is not yet sorted out. TPP, Trans-Pacific Partnership.

A number of community farmers were rescued by the government subsidy for rice for animal feed in 2014, when the rice price fell sharply. Hamlet G in Tsuwano-chō, Shimane prefecture, was about to establish community farming when the rice price dropped drastically in 2014. Because producing rice for meals would only result in a deficit for the operation's income and expenditure plan, some farmers opposed launching community farming. Hamlet G changed plans and proposed growing rice for animal feed instead to make a profit. And, thus, Hamlet G began community farming.

Rice for animal feed plays a major role not only in wet paddies but also in those paddies that can produce wheat and soybeans. Some farmers tried a method that was much talked about around the second turning point in 1978: rotating rice and other upland field crops at regular intervals with rice for animal feed instead of rice for meals.

Some other farmers who owned rice paddies unsuitable for upland field crops decided to convert from growing soybeans to growing rice for animal feed. In one case, friction arose between two farmers who had been growing soybeans, which are vulnerable to dampness, in adjacent rice paddies when one of them converted from soybeans to rice for animal feed in hopes of obtaining the government subsidy. Even in regions known as major producers of well-known brand rice, there were farmers who chose to grow rice for animal feed because of the high government subsidy. How to produce rice for animal feed differs significantly depending on the farming region, the rice paddy, and the attitude of rice farmers.

Table 5.2. Amounts of rice eaten by humans and livestock. *Source*: based on the Food and Agriculture Organization Statistics (FAOSTAT) database of the United Nations. *Note*: anything below 1000 t is rounded down.

	Comparison of amounts eaten (ratio of humans:animals)		Growth rate in 50 years	
	1963	2013	Amount of rice eaten by humans	Amount of rice eaten by livestock
World	22:1	11:1	2.8 times ↑	5.6 times ↑
Europe	21:1	9:1	2.1 times ↑	4.8 times ↑
North and Central Americas	1:0	1:0	4.9 times ↑	–
South America	452:1	30:1	2.9 times ↑	43.6 times ↑
Asia	22:1	10:1	2.7 times ↑	5.7 times ↑
Africa	15:1	53:1	8.5 times ↑	2.4 times ↑
Japan	385:1	75:1	0.7 times ↓	3.6 times ↑

When the Japanese Started to Use Rice as Animal Feed

The production of rice for animal feed is increasing

Rice is predominantly eaten by people. This fact makes it difficult to know exactly by how much, but the amount of rice fed to animals has been increasing steadily all over the world (see (Table 5.2) . While the amount of rice eaten by people all over the world rose by 2.8 times in the half century between 1963 and 2010, the amount eaten by animals increased by 5.6 times.

As exemplified by North and Central America, where rice is not fed to animals, and Africa, where rice is eaten chiefly by people, use of rice as animal feed differs region by region. Japan is in a unique situation where the amount of rice eaten by humans is decreasing while the amount eaten by animals is increasing.

Rice for animal feed as a solution to the rice surplus problem

Because the rice price in Japan is higher than the global standard, all the rice for animal feed, excepting scrap rice, is subsidized by the government. And the method to make rice into animal feed is not confined to diverting surplus rice for use as animal feed or growing rice for animal feed. Minimum access rice, for instance, is also made into animal feed.

Until fiscal year 2005, minimum access rice, which is imported to fulfill Japan's World Trade Organization quota, was used as (i) rice mainly for use in the restaurant industry, (ii) rice to be processed into miso and soy sauce, and (iii) rice for overseas food aid. When the supply of minimum access rice became overstocked, it was decided to use it as animal feed from 2006. After November 2005, the starting month of the statistics for 2006, 54% of minimum access rice was used as animal feed (see Fig. 5.11). In recent years, as much as 70–80% of imported minimum access rice has been used as animal feed.

Another example is stockpile rice: it is made into animal feed when it gets old. It used to be that the national government would store its stockpile rice for a year or two before selling it as rice for meals. Since 2011, however, stockpile rice has been separated out from rice for meals, eliminating its influence on the price of rice for meals. Subsequently, rice entering Japan under the TPP agreement, which can be categorized as imported rice, came to be procured by the government as stockpile rice. Rice in the government stockpile is stored for 5 years. Unless it is released in cases of emergency, consumers almost never cross paths with it in day-to-day life. In the end, almost all of the stockpile rice is used as animal feed (see Fig. 5.12).

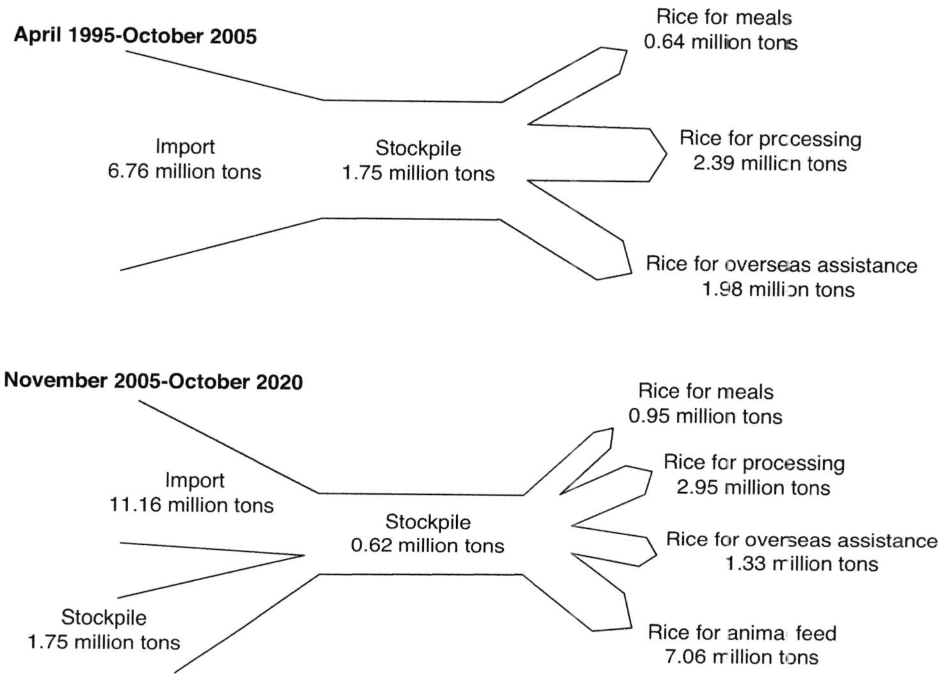

Fig. 5.11. Situations of export, stockpile, and sale of minimum access rice. *Source*: based on the Ministry of Agriculture, Forestry and Fisheries' *Kome wo Meguru Jōkyō ni Tsuite* [*On Situations Surrounding Rice*]. *Note*: excluding 0.2 million tons of rice that were specially processed and sold.

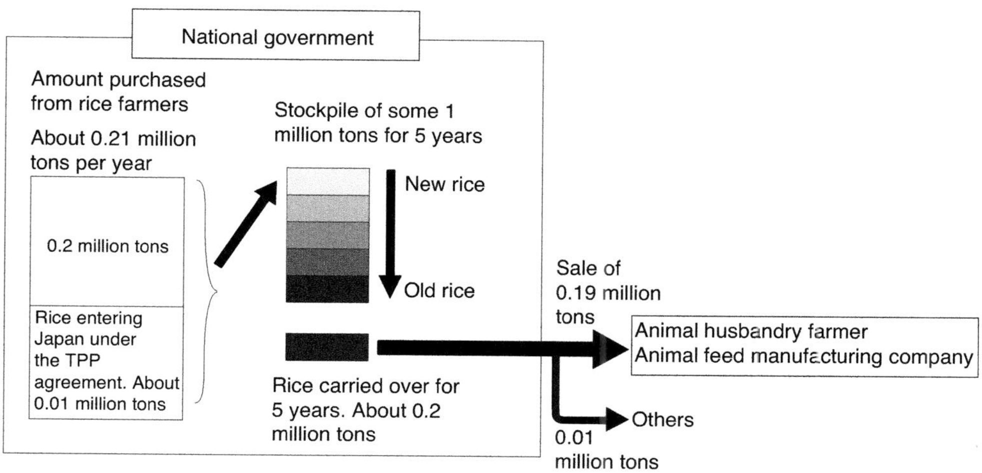

Fig. 5.12. Mechanism for stockpiling rice. *Source*: based on the Ministry of Agriculture, Forestry and Fisheries' *Kome wo Meguru Jōkyō ni Tsuite* [*On Situations Surrounding Rice*] and *Sōgōteki na TPP-tō Kanren Seisaku ni Motozuku Nōrin Suisan Bun'ya no Taisaku* [*Measures of the Agriculture, Forestry, and Fisheries Sectors Based on Outline of Comprehensive Policies Related to the Trans-Pacific Partnership*].

Under the current policy, a rice farmer is allowed to sell surplus rice from 1 year's good harvest in the next year if he increases production of rice for animal feed by the same amount in that next year.

Additionally, there exists a mechanism that would be utterly unbelievable for vegetable farmers. When the national government predicts a rice surplus resulting from a better-than-average harvest, it induces rice farmers to designate the rice they are growing for meals as rice for animal feed. By changing the name of the crop, systems and subsidies are maneuvered. Although this is an AD hoc measure, it is a fact that in 2020 rice for meals being produced in 4000 ha of paddy fields was converted to rice for animal feed.

Reasons why sacred rice became animal feed

At the annual *Niinamesai*, a harvest ritual, the emperor of Japan thanks the Shinto deities for a prosperous year and prays for a fruitful new year. All over Japan, people give thanks for the year's harvest of rice. In some families, parents teach their children that a god dwells in rice. Since ancient times it has been taught that animals are inferior to humans and, thus, feeding rice to animals is criticized as going against Japanese sensibilities.

The sudden increase in the production of rice for animal feed is a reflection of a new attitude on the part of the Japanese. This new attitude may appear to contradict the Japanese people's traditional sanctification of rice. Instead of finding here a discontinuity in the Japanese people's spirituality regarding rice, however, I myself detect consistency in the perception of rice production being something that is given to them. I believe all Japanese can share this sentiment at the fundamental level.

According to the *Nihon Shoki* (*Chronicles of Japan*), finished in 720 AD, upon the descent to Earth of the grandson of the sun goddess, people were given rice that Amaterasu Ōmikami (goddess of the sun) grew in a rice paddy in the High Plain of Heaven (home of the gods). This story, in short, says that Japan's rice was god-sent. In modern times we do not have to resort to mythology to point out that rice was given to all areas of Japan, including those without any rice culture, through a supply system known as the rice-rationing system. For the Japanese people, rice has always been something given to them.

On the other hand, it should be noted that the rapid proliferation of farmers growing rice for animal feed was made possible by government subsidy. Rice farmers did not make any effort or sacrifice to obtain a subsidy for growing rice for animal feed. It was the national government that created rice for animal feed. From the viewpoint of rice farmers, rice for animal feed was given to them with some policy purpose attached.

The history behind the birth of rice for animal feed includes something intangible yet not entirely detached from reality, such as the need for food security. Just like the mythology of rice being given to people by a god, the origin of rice for animal feed has some element which seems far from reality but somehow contiguous with it.

Although it is rice farmers who produce rice for animal feed, they did not conceive it. Rice for animal feed, which would not exist without government subsidy, is rice given to them from beginning to end. Had the subsidy for rice for animal feed been a result of the rise of the Rice for Animal Feed movement, which was a farmers' movement, or a consequence of pressure coming from people's awareness of the need for food security, the rapid increase in the production of rice for animal feed would must have brought with it a major disconnect from or a psychological conflict over the Japanese people's relations with rice.

Some scholars attribute the rise in the production of rice for animal feed to the generational change in Japan from those born during the first decade of the Shōwa period (1926–1934), who had an aversion to rice for animal feed, to their successors, who pay no it heed. It is my view, however, that the way rice for animal feed was introduced – with endorsement by the state and justifications and massive subsidies attached – is the greatest reason rice farmers had no problem producing this rice. The coexistence of sanctified rice and rice to be fed to animals is accepted by the Japanese people as consistent in the fact that, as they see it, rice is given to them.

Saiden Nukiho no Gi, the ceremonial harvesting of rice to be used in *Daijosai*, the first ceremonial offering of rice by a newly enthroned Emperor, in Takanezawa-machi, Tochigi prefecture (*The Asahi Shimbun*, September 27, 2019). (Used with permission from Asahi Shimbun)

Importance of Japan's animal husbandry business

To digress a little from the issues of rice and rice paddies momentarily, let me point out that today's livestock are nurtured with feed on a par with human food, and livestock eat a lot more than humans. If ever Japan was to face another war-induced food shortage, therefore, it will have to be beefsteak, not a simple meal of rice and side dishes of vegetables, that should be served first in order to prevent corn, the main staple for livestock, from being consumed by animals. Livestock products are sumptuous food indeed. But it requires 11 kg of cereal to produce 1 kg of beef, 7 kg of cereal in the case of pork, and 4 kg in the case of chicken. As much as 180g of cereal are required to produce one egg. On top of that, producing animal feed requires a tremendous amount of land and water. While the world has seen an acceleration in research and development and the commercialization of meat alternatives, insects as food, and artificially cultured meat, a production method requiring massive inputs of resources still remains the mainstream in today's animal husbandry industry.

The animal husbandry industry also plays a role as an outlet for leftover food. In retrospect, we note that Japan's animal husbandry industry, which was developed after the Pacific War, once was an outlet for surplus cereal from the United States. As such, the industry in Japan contributed to promotion of friendly relations between Japan and the United States as well as to mitigating bilateral trade friction. The industry has also provided an AD hoc solution to the surplus rice problem and a measure to prevent imported rice and stockpile rice from competing with rice for meals Moreover, Japan's animal husbandry industry has contributed to the effective utilization of surplus rice paddies through the production of FWCS and rice for animal feed, as well as to Japan's food security in case of emergency.

Seventy percent of the food waste, including rice, disposed of by the food industry – or 20 million tons annually – is reused. The majority of that (70%) is used as animal feed. The amount of animal feed that originated as food waste increases year by year. Today, food waste reused as animal feed amounts to 10 million tons, which is equivalent to 1.5 million tons of corn, or 15% of the corn imported annually. Here let me point out that the animal husbandry industry continues to have an important role in the Japanese economy and Japan's security that is irreplaceable by any other industry.

The Flipside of the Full Utilization of Rice Paddies

Reasons behind the high subsidy for rice for animal feed

The size of the government subsidy for surplus rice paddies fluctuates year by year, but compared with what it was in the early 1980s, it has remained high. In fiscal year 2021, the amount reached a record high (see Fig. 5.13). However, as one can detect from the comparison of the subsidy per 10 ares of surplus rice paddy to the starting salary of a college graduate in the period between the early 1990s to 2020, the amount in 2021 was not so large in today's wage terms.

The government subsidy for surplus rice paddies peaked in 1979, 1 year after the second turning point. This was followed by years of smaller subsidies per hectare. In fiscal 2021, the

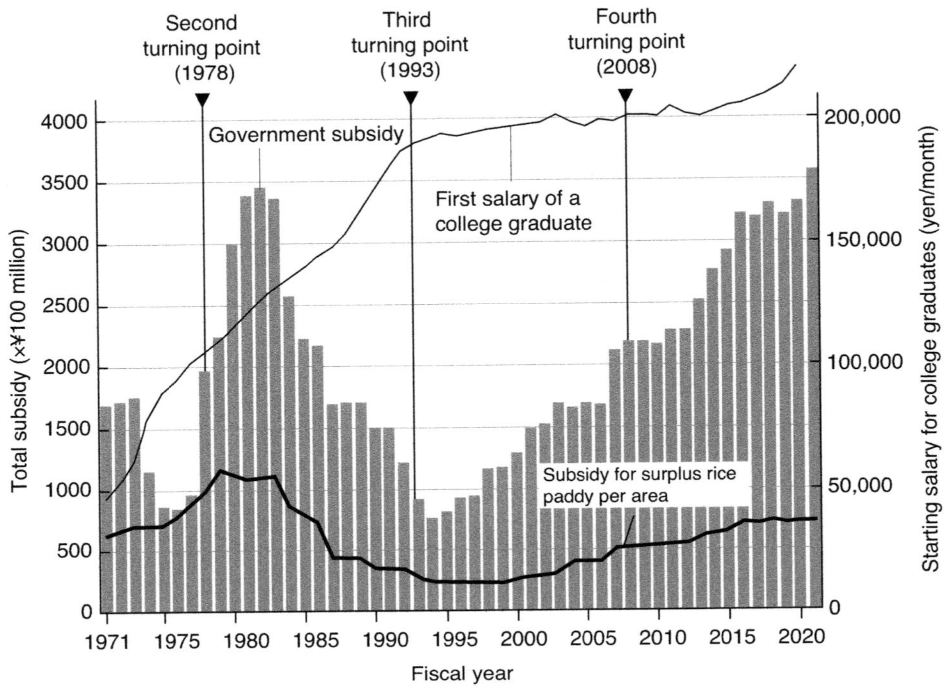

Fig. 5.13. Changes in the subsidy for surplus rice paddy area and first salary of a college graduate. *Source*: based on the Ministry of Agriculture, Forestry, and Fisheries' *Sakumotsu Tōkei* [*Statistical Survey on Crops*], Japanese government budgets for various fiscal years (initial and supplementary, formulated in each same fiscal year, with some of the supplementary budgets possible application to production in the ensuing year), and the Ministry of Health, Labor and Welfare's *Chingin Kōzō Kihon Tōkei Chōsa* [*Basic Survey on Wage Structure*].

subsidy was ¥35,787 per 10 ares, or ¥357,870 per ha. The government nevertheless continued to provide ¥550,000 to ¥1.05 million per ha subsidy for rice for animal feed on the condition that the subsidy be used exclusively for maintaining those rice paddies that produced rice for animal feed. Some readers may be under the impression that the subsidy for surplus rice paddies was provided for all surplus paddies (Fig. 5.14, right-hand side), but that was not the case (see Fig. 5.14, left-hand side). For instance, in fiscal year 2018, only 67% of surplus rice paddies benefitted from that year's ¥298.6 billion subsidy for full utilization of rice paddies. Moreover, although ¥77.2 billion or 26% of the entire subsidy was poured into growing rice for animal feed, only 9% of the total surplus rice paddy area was actually used.

The total area of surplus rice paddies nationwide includes paddies not used to produce rice for meals at all; these paddies receive either a smaller subsidy or no subsidy. Furthermore, some rice paddies are going to disappear. The high subsidy for rice for animal feed has been made possible by the presence of these rice paddies.

Twilight of full utilization of rice paddies?

While the official reason for the policy on the full utilization of rice paddies is to promote the use of surplus rice paddies, there is a hidden aspect, which is that full utilization is realized because surplus rice paddies are not used. Let us sort

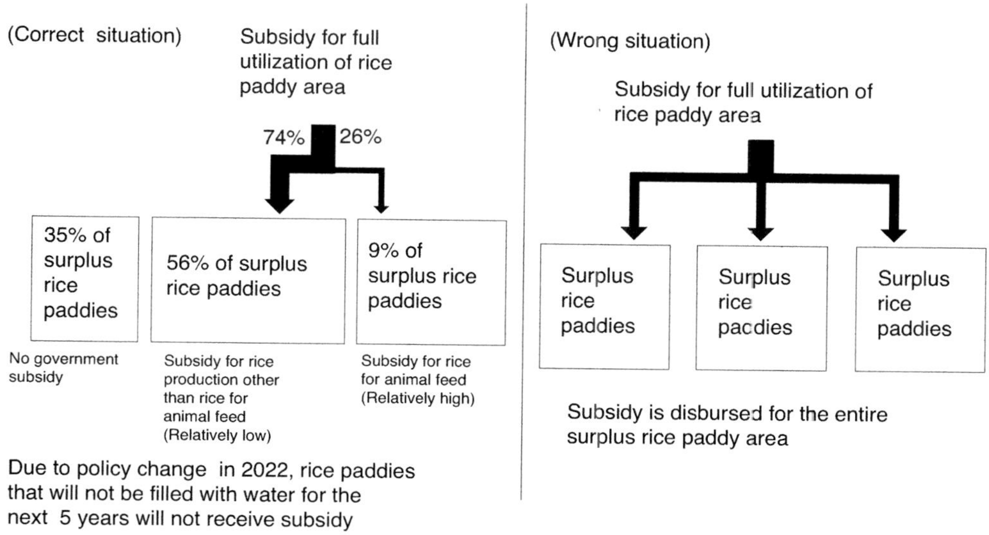

Fig. 5.14. Correct and incorrect illustrations of the subsidy for full utilization of rice paddy area, 2018. Left: the correct situation; right: the incorrect, or wrong, impression. *Source*: based on the Ministry of Agriculture, Forestry, and Fisheries' *Sakumotsu Tōkei* [*Statistical Survey on Crops*] and Finance Ministry documents on the fiscal year 2020 government budget proposal. *Note*: double cropping and collaboration between crop cultivation farms and livestock farms are excluded.

this out from the angle of changes in the three meanings of rice paddy full utilization.

Full utilization of rice paddies was a campaign borne of the strife between the LDP and the DPJ. Until 2013, three meanings of the phrase held steadfast: (i) as a synonym for the policy to curb rice production, (ii) as an effort to make as wide a use of rice paddies as possible, and (iii) as an effort to grow as much agricultural produce as possible. In 2014 and the years following, the meaning of full utilization of rice paddies underwent a major change. When the resumption of rice policy reform took the spotlight, the first meaning of the phrase as a synonym of the policy to curb rice production was highlighted at the expense of the effort to make as wide use of rice paddies as possible. This situation continues to date.

Typical among the changes is the position of rice for animal feed. Immediately after the fourth turning point in 2008, the LDP proposed that rice for animal feed should be produced in areas or rice paddies that were not suitable for wheat or soybeans. Production of rice for animal feed was positioned as a means to utilize rice paddies where nothing was planted. It was supposed that rice for animal feed would be grown in borderline surplus rice paddies whose yield of agricultural produce was questionable. Because the aspect of full utilization of rice paddies as a synonym for the policy to curb rice production came to be stressed over the two other meanings around 2014, introduction of rice for animal feed to those rice paddies no longer used to grow rice for meals was promoted (see Fig. 5.15).

As a result of this change in the position of rice for animal feed, a mechanism was introduced in 2014 to make it easy for rice farmers to give up on producing rice for animal feed in borderline rice paddies. Under the renewed institutional design, paddies whose yield of rice for animal feed had been lower than other paddies in a region received a smaller subsidy.

In addition, the third meaning of full utilization of rice paddies as an effort to produce as much agricultural produce as possible was also overshadowed. The subsidy to promote double cropping, started by the government in 2010, was changed in 2017 to a measure to be pursued voluntarily by each region. The result was a de facto reduction in the government subsidy. In order to reduce production of rice for meals, it

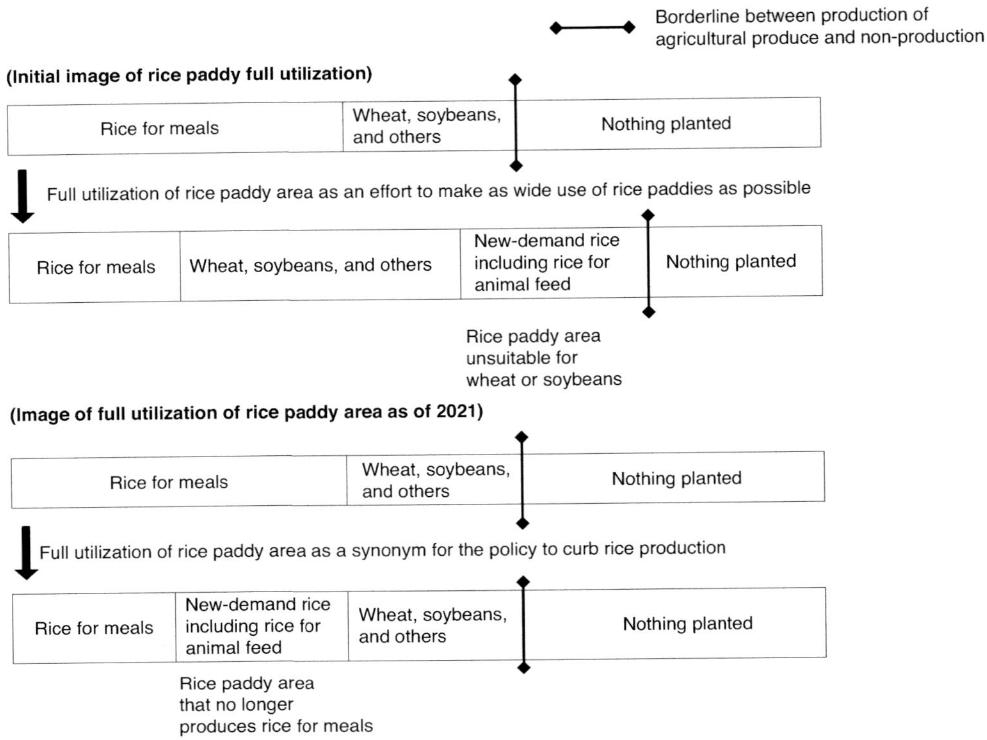

Fig. 5.15. Changes in the full utilization of rice paddy area.

was found to be more effective to concentrate the subsidy on summer produce that was in competition with rice for meals than it was to subsidize double cropping, because double cropping means paddies have to be used in wintertime.

The LDP advocated permanent securement of a budget for full utilization of rice paddies in general elections in 2017, 2019, and 2021. As such, the government's rice paddy policy continued to be centered around full utilization of rice paddies. Still, the substance was different from the rice paddy full utilization as first proposed by the LDP. In fact, the substance was closer to the rice policy reform in 2004–2007. Although the measure to handle the rice paddy surplus problem was reinforced by the fourth turning point in 2008, the problem ended up remaining unsolved and the government returned to a rice-centered policy again.

Today, in prefectures and regions comprised of both individual cities and towns and municipalities formed as collectives of cities and towns, local councils composed of local government representatives and agricultural associations envision the future prospects of rice paddies. In 2021, the wording 'full utilization of rice paddies' that had been part of the 'vision on full utilization of rice paddies' discussed at these councils was abandoned in favor of the wording 'vision on improvement of rice paddies' earning capacity.' Furthermore, on the national government's initiative, the vision's focus shifted from the prospects of rice paddies, including borderline rice paddies, to the prospects for expansion of profitable agriculture.

The weakened measure for unplanted dry farmland and strengthened drive to reduce rice paddy area

Looking closely at the visions outlined in various regions immediately before and after abolishment of the rice paddy reduction program in 2018, we see that by fiscal year 2017 prefectures and multi-prefecture regions had included the goal of 'resolution of unplanted dry farmland' and worked out measures toward achieving this goal. As far as the documents I have managed to obtain are concerned, 33 prefectures out of 35 recorded their own measures toward this end in 2017, of which 21 announced that new-demand rice would be produced in the unplanted dry farmland. When 'resolution of unplanted dry farmland' disappeared from the national government program in 2018, only four out of 39 prefectures put down measures for 'resolution of unplanted farmland.'

On the other hand, in fiscal year 2017 none of the 35 prefectures included 'promotion of expansion of cultivated fields' as a goal in their visions. When the national government encouraged 'promotion of expansion of cultivated fields' in fiscal year 2018, however, 33 prefectures out of 39 named that as a goal. In fiscal year 2021, the national government strongly urged prefectures to include 'policies and goals as a producer of agricultural produce toward effective use of rice paddies, including expansion of cultivated fields' in their visions. This government stance has promoted the expansion of cultivated fields in various regions.

Unlike producing upland dry field crops in rice paddies, converting rice paddies to cultivated fields means farmers lose the right to receive government subsidies for surplus rice paddies. The subsidy offered for converting paddies to cultivated fields was, so to speak, consolation money. More concretely, a new discretionary subsidy of ¥1.05 million per ha that had been instituted in 2018 was converted in 2020 to a subsidy directly disbursed by the national government. In 2021, this subsidy was raised to ¥1.75 million per ha.

As rice paddies disappear, subsidies for surplus rice paddies can be used for smaller areas of farmland. Presently, the subsidy for rice paddy reduction is disbursed as part of the subsidy for surplus rice paddies. In that sense, this subsidy is part of the policy for full utilization of rice paddies. The reality that full utilization of rice paddies is still being advocated officially while rice paddy reduction also is being pursued reveals that the rice paddy surplus problem remains unsolved even after the fourth turning point in 2008.

Policy change in 2022

Triggered by a November 2021 proposal by the government and the LDP, a new rule was suddenly made that, starting in fiscal year 2022, rice paddies that had not been filled with water during the next 5 years were no longer eligible for the subsidy for surplus rice paddies. There already had been a rule that land which could hardly be called a rice paddy was ineligible for the subsidy issued under the rice paddy reduction measure. But this new rule was an extraordinary development in that it established a stringent regulation for a specific 5 year period.[3]

The 2022 rule not only terminates subsidies for highly profitable produce, which without the subsidy would not have been grown in rice paddies, but also leads to the cutoff of unused rice paddies and rice paddies with low profitability that had relied on the subsidy to continue growing upland dry field crops. Because there is no openly available statistical data on any rice paddy's history of being filled with water, the exact area of rice paddies affected by this new rule is unknown. Undoubtedly, vast numbers of rice paddies are affected. Especially those surplus rice paddies whose owners, after sincere contemplation, had decided on and had been steadfastly implementing usage of paddies without relying on rice – particularly, those rice paddies where such low-profit crops as pasture, wheat, and soybeans have been continuously produced – are gravely affected. The meaning of full utilization of rice paddies as a synonym for the policy to curb rice production has intensified, being far removed from the initial meaning of the proposition.

The content of the policy to curb rice production easily fluctuates. With the sudden changes to the policy in 2022, it remains liable to be abruptly cancelled or de facto watered down through a relaxation of standards. If this new rule persists, the attention that the problem of surplus rice paddies has received since the fourth turning point in 2008 would be put to rest, decisively marking the end of the era of full utilization of rice paddies.

The 'end' here does not mean the arrival of another, new turning point in 2022. What the end of the era of full utilization of rice paddies means is, simply, continuation of the rice paddy surplus problem and reversal of the previous, rice-centered orientation. Therefore, if the 2022 policy change rule continues, people in the future would not perceive 2008 as a major turning point. As for merits and demerits of the 2022 policy change concerning the filling of rice paddies with water, these will be discussed in detail in Chapter 6, where we will explore further.

The Year 2008 in Retrospect

The reform attempt at controlling rice paddy usage through production of marketable rice and nurturing rice farmers specializing in marketable rice was crippled shortly after it was launched. While the reform attempt entered the spotlight due to rapid falls in the rice price and the DPJ's surge of popularity in national politics, its failure was actually attributable to neglect of the rice paddy surplus problem.

With 2008 as the turning point, a system to control rice production by expanding subsidies for surplus rice paddies was adopted. This new system, nurtured and refined again and again by both the LDP and the DPJ throughout their rivalry over the course of two changes of government, was strengthened as a measure for handling the rice paddy surplus problem. The government subsidy for the production of rice for animal feed was representative of this.

The measure to address the rice paddy surplus since 2008 is called the promotion of full utilization of rice paddies. The phrase had three meanings as (i) a synonym for the policy to curb rice production, (ii) an effort to make as wide a use as possible of rice paddies, and (iii) an effort to produce as much agricultural produce as possible. Over time, the balance among the three meanings changed. Today, the measure heavily inclines toward the first meaning as a synonym for the policy to curb rice production.

As a result, while the notion of full utilization of rice paddies that emerged at the fourth turning point of 2008 had the effect of acting as an excuse to increase government subsidies to curb production of rice for meals, it came short of solving the rice paddy surplus problem. In an attempt to mitigate the pain farmers felt because of the rice paddy surplus problem, consolation money has been given to rice farmers who reduced their rice paddy area. And yet, the fundamental solution to the problem of surplus rice paddies has not been found. In other words, the rice paddy surplus is a gigantic problem that remains unsolved to date.

Nevertheless, this unsolved problem is nothing but a canary in a coal mine heralding the greatest crisis still to come for the Japanese people. When we look back at the four turning points that rice and the Japanese people have gone through, paying special attention to the issue of the rice paddy surplus, as we will do in the next chapter, another problem will be illuminated that could lead to the greatest danger for Japan and its people.

Notes

1 For the sake of simplicity, political parties that played central roles, such as the Liberal Democratic Party and the Democratic Party of Japan, are used to explain actions taken by political parties instead of coalition governments or parliamentary groups.

2 'Beika wo Otte 2' [Following the rice price, 2], *Yomiuri Shimbun*, September 29, 1967.

3 After its initial publication, the new rule, which was created by a policy change in 2022, was effectively withdrawn as of April 1, 2025, after Minister of Agriculture, Forestry and Fisheries Taku Eto indicated on January 31, 2025 that he would review the rule. This was a sudden turn of events. These events occurred after the original book was written. This book was written based on the situation before the sudden change in policy.

References

Kumagai, H. and Ohtani, T. (eds) (2009) *Shiryōmai no Seisan to Buta Nikushitsu no Kōjō* [*Production of Rice for Animal Feed and Improvement of Pork Meat Quality*]. Nōrin Tōkei Shuppan, Tokyo.

Ozawa, W. and Yoshida, N. (2009) *Shiryōyō Mai no Saibai/Riyō* [*Harvest and Use of Rice for Animal Feed*]. Sōshinsha, Tokyo.

Taniguchi, N., Lee, Y., Umemoto, M. and Senda, M. (2010) *Suiden Katsuyō Shinjidai* [*New Age of Rice Paddy Use*]. Nōsan Gyoson Bunka Kyōkai, Toda.

6 At the Root of Today's Rice Problem

Abstract

Chapter 6 summarizes the discussions presented in Chapters 2–5 as well as the characteristics of today's rice problem, analyzing changes in the economic size of the rice industry and the history of wet paddies. The chapter first points out that the problem caused by the rice surplus confronted at the first turning point was replaced by that of excess rice paddies. Second, most of the apprehension about excess rice paddies was replaced by worry over the increasing fragility of food security and the multifunctional role of agriculture and rural areas. Citing budgetary constraints and the lack of transparency in policy management, Chapter 6 makes the third point that the issue of how to institutionally treat non-rice-producing rice paddies remains unsettled.

The Current Status of the Japanese People, Rice, and Rice Paddies

Agriculture's departure from rice

This chapter will review the four turning points Japan has experienced regarding rice. We will explore the rice problem more deeply by looking at the history and problems of rice and rice paddies from various angles.

First, let us confirm the changes that occurred in rice production between the first turning point in 1967 and 2020 (see Fig. 6.1). The amount of rice eaten during this period decreased by 41% from 10.36 million tons per year to 6.1 million tons. The share of rice in total agricultural output dwindled from 45.5 to 18.5%, or less than half of the traditional share. In the same period, the number of rice farmers decreased by as much as 85% from 4.88 million to 0.71 million, while rice paddies where rice for meals was produced dropped by 57% from 3.15 million ha to 1.37 million ha.

These facts refute the popular belief that Japan's agriculture and farmers have persisted with rice. Detachment from rice by farmers and the agricultural industry has advanced at a faster pace than the Japanese people's loss of interest in eating rice.

Focusing only on rice paddies, though, tells a totally different story. Admittedly, rice paddy area over the same period of 1967–2020 did diminish by 29%. But the rate of decline was smaller than the rate of people's loss of interest in eating rice. In other words, it is rice paddies, not rice per se, that Japan's agriculture has persisted with. And the symbol of this obsession is surplus rice paddies: they occupy some 40% of total rice paddy area. The day when it can be taken for granted that rice and rice paddies are connected is already long gone.

Changing connection between rice and rice paddies

Let us revisit the four turning points through the lens of changes in the connection between rice and rice paddies over time. In the past Japan has experienced rice exports and deserted cultivated land, as it is called today, but it is the megatrend illustrated in Fig. 6.2 that we concentrate on here.

For the Japanese people, rice and rice paddies have been inseparable since ancient times. Rice paddies are cultivated in order to harvest rice, and rice is grown in rice paddies. While there were times when Japan had to rely on imported rice or rice produced in Taiwan and Korea due to a shortage of rice paddies, the connection between rice and rice paddies remained unchanged.

The first turning point in 1967 when self-sufficiency in rice was attained also marked the time when the interrelation between rice and rice paddies was no longer taken for granted.

DOI: 10.1079/9781800627093.0006

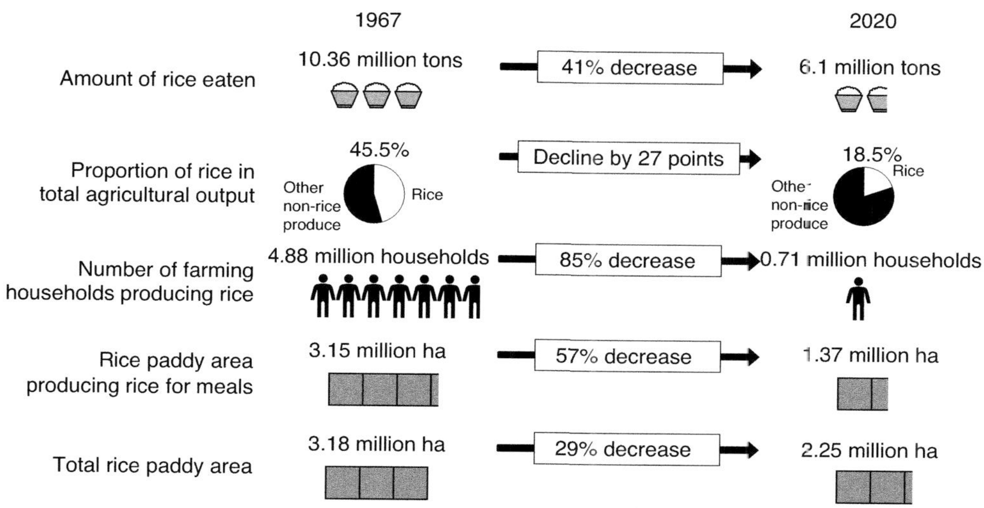

Fig. 6.1. Japanese people's detachment from rice and rice paddies. *Source*: based on the Ministry of Agriculture, Forestry and Fisheries' *Sakumotsu Tōkei* [*Statistical Survey on Crops*], *Nōgyō Keiei Tōkei Chōsa* [*Statistical Survey on Farm Management and Economy*], and *Shokuryō Jukyūhyō* [*Food Supply and Demand Balance Sheet*]. *Note*: numbers of farming households represent the number of total farm households in 1965 and the number of agricultural business organizations taken from the Census of Agriculture and Forestry in Japan conducted every 5 years by the agriculture ministry.

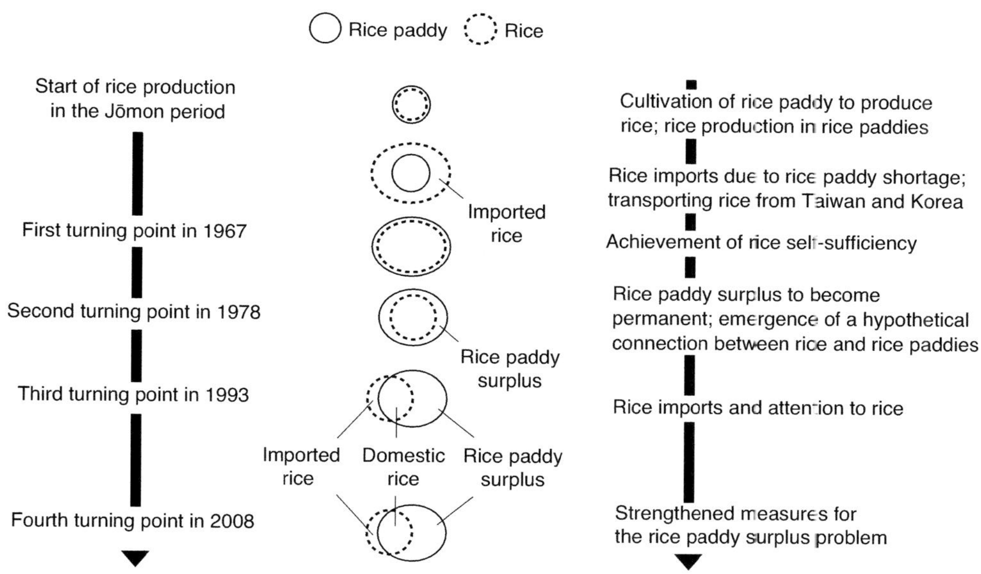

Fig. 6.2. Changes over time of the relationship between rice and rice paddies.

At this turning point, the curtain rose on the era when the amount of rice produced would be in excess of demand if rice was grown in all rice paddies. Subsequently, a mechanism replacing the rice surplus problem with the rice paddy surplus problem was consolidated by the rice paddy usage reorganization measure at the second turning point in 1978. Around this time, people's concern about food security had heightened, leading to a hypothetical connection between rice and rice paddies which said that surplus rice paddies could be used to produce rice in times of emergency.

At the third turning point in 1993, when the Uruguay Round was in essence concluded, rice imports began when rice paddy area was in surplus. This was quite a different situation from the time before the first turning point. Furthermore, rice distribution was liberated from state control in compliance with international rules of agricultural policy, and the government shifted its tactics for controlling rice paddy area by pursuing a rice-centered ideal vision of rice policy.

Although concern about rice paddies continued to mount through the first to third turning points, it did not lead to a fundamental solution of the rice paddy surplus problem. At the occasion of the fourth turning point in 2008, the idea to tighten control of the production of rice for meals by strengthening measures targeting the rice paddy surplus, in addition to pursuing a rice-centered ideal vision of rice policy, became stronger.

History of policy to curb rice production

Next let us review the history of the policy to curb rice production, taking the four turning points into consideration.

Originally, the policy to curb rice production began as a state request to rice famers for their cooperation. Thus, to the question, 'When will government coercion regarding the reduction of rice production end?", the correct answer is, 'It never even started.' When the policy was first launched, there was not much surplus rice paddy area and the government induced rice farmers' cooperation using high per-area subsidies as a 'carrot' as well as through some 'stick' measures.

Even though the policy took the form of a request from the state, the stick accompanying the request became increasingly austere after the second turning point in 1978. The policy was enforced through heavier reliance on appeals from local governments and agricultural cooperatives as well as on peer pressure among rice farmers themselves. As surplus rice paddy area continued to increase, the government succeeded in dissuading farmers from producing rice with its carrot-and-stick approach that carried an air of coercion. In this way, it minimized its financial burden. Still, the shortage of subsidies for allowing production of agricultural produce in all surplus rice paddies persists to date.

As the result of abolishment of the Staple Food Control Act at the time of the third turning point in 1993, the national government pursued the policy to curb rice production not by strong request for cooperation, but by stressing its direct benefits for rice farmers themselves. At this point, the government subsidy for surplus rice paddy area was not abolished because it had played a role in the public interest. Had the subsidy been eliminated, the use of surplus rice paddies could have changed drastically. If, for instance, production of rice for meals in surplus rice paddies increased rapidly, the rice price would fall sharply. This would make rice production highly volatile and, consequently, threaten people's livelihood and the stability of Japan's economy as a whole. If rice paddies are left to deteriorate, they become useless in times of emergency, and it also becomes impossible for Japan's agriculture to realize its multifunctional roles.

While the government subsidy for surplus rice paddy area has not been sufficient over the years, it did begin to be increased after the fourth turning point in 2008. Subsequently, carrot-and-stick measures aimed at rice farmers who refrained from growing rice were terminated. Instead, it was re-emphasized that rice farmers and agricultural organizations should curb rice production for their own benefit. Also, the government inclined more toward controlling production of rice for meals by providing government subsidies for surplus rice paddy area.

As rice was transformed from military goods to a general commodity, the government's appreciation of how surplus rice paddies can serve the public interest remained high. And this became the greatest reason for the delay in reducing surplus rice paddy area. At the same time, detachment from rice as a business quickly advanced.

Turning point as detected by the assessed value of rice

Let us now review the changes in the assessed value of rice (the rice price multiplied by the amount produced) after the second turning point (see Fig. 6.3). Because the rice price varies by species and brand, we will use the value that allows comparison by year, so as to minimize the influence of changes in consumer prices.

Comparing the price at which the state bought rice in the year before each of the first and third turning points (1978 and 1993), we see that the assessed value of rice declined from ¥5465.5 billion to ¥3058.6 billion. This decline is attributable to, first, the decreased amount of rice overall (accountable for ¥580 billion of the total decline); second, lower rice prices (accountable for ¥1535.6 billion); and third, the effects of both the amount of rice and price declines (accountable for ¥291.2 billion).

The nominal price of rice per 5 kg at this point dropped from ¥1438 in 1978 to ¥1366 in 1992, a decline of merely 5%. As a means to cope with the widening income gap between the agriculture and manufacturing sectors, the nominal price of rice was kept unchanged or at least limited to a small drop. As general consumer prices rose, however, the real price of rice fell, too.

When, at the third turning point in 1993, the state stopped supporting the rice price, it dropped quickly. Comparing the rice price in 1994, the year after the historic poor harvest in 1993 that had caused a stunning rise in the rice price, with the price of rice traded in 2007, 1 year before the fourth turning point in 2008, we find that rice's assessed value declined from ¥4627.1 billion to ¥2181.5 billion under the strong influence of smaller amounts of rice being harvested and falls in price. The government subsidy introduced in 1998 was meant to compensate for the drop in price that had accompanied the third turning point in 1993, but its amount was not enough to cover the entire price drop and the subsidy was exclusively for those who complied with the policy to curb rice production. For rice farmers at the time of the July 2007 House of Councillors election, which triggered the fourth turning point in 2008, the falling price of rice, the dwindling amount of rice production and the resultant increase in surplus rice paddy area were equally important matters of concern.

Subsequently, from the fourth turning point in 2008 to 2020, the assessed value of rice declined from ¥2284.3 billion to ¥1749.9 billion. This was more attributable to the reduction in the amount of rice produced than to drops in the rice price. Due to the decline in the amount of rice produced, the assessed value of rice decreased as much as ¥346.1 billion. This amount was almost equal to the government subsidy for surplus rice paddy area in 2021, which was ¥346.0 billion (the sum of the regular budget and the supplementary budget). While the stagnant rice price has been a source of worry for today's rice farmers, it is far more worrisome that reductions in rice production may lead to a decrease in income from rice production. And the major concern for rice farmers is whether they can make up for dwindling income with their surplus rice paddy area or not. Compensation for reduced income with subsidies for surplus rice paddies, in particular, greatly affects the business of rice farmers.

Wet Paddies: Where Rice Paddies and Rice Are Strongly Linked

What are wet paddies?

Today, when the connection between rice paddies and rice cannot be taken for granted, the strong link between wet paddies, where no crop other than rice can be easily grown, and rice tends to be forgotten. In this section, let us consider wet paddies in order to shed light on the current status of the rice problem.

Wet paddies are rice paddies with poor drainage. Depending on their possible usages, however, wet paddies have been defined in various ways; for example, wet paddies can be defined as paddies hard for an army to march over, paddies difficult to plow using oxen, horses, or machines, and paddies where it is not easy to grow dry upland crops.

Today's maps use only one icon to indicate rice paddies, but a 1917 map used three icons to differentiate dry paddies, wet paddies, and deep-mud paddies. In demarcating the three, the Land

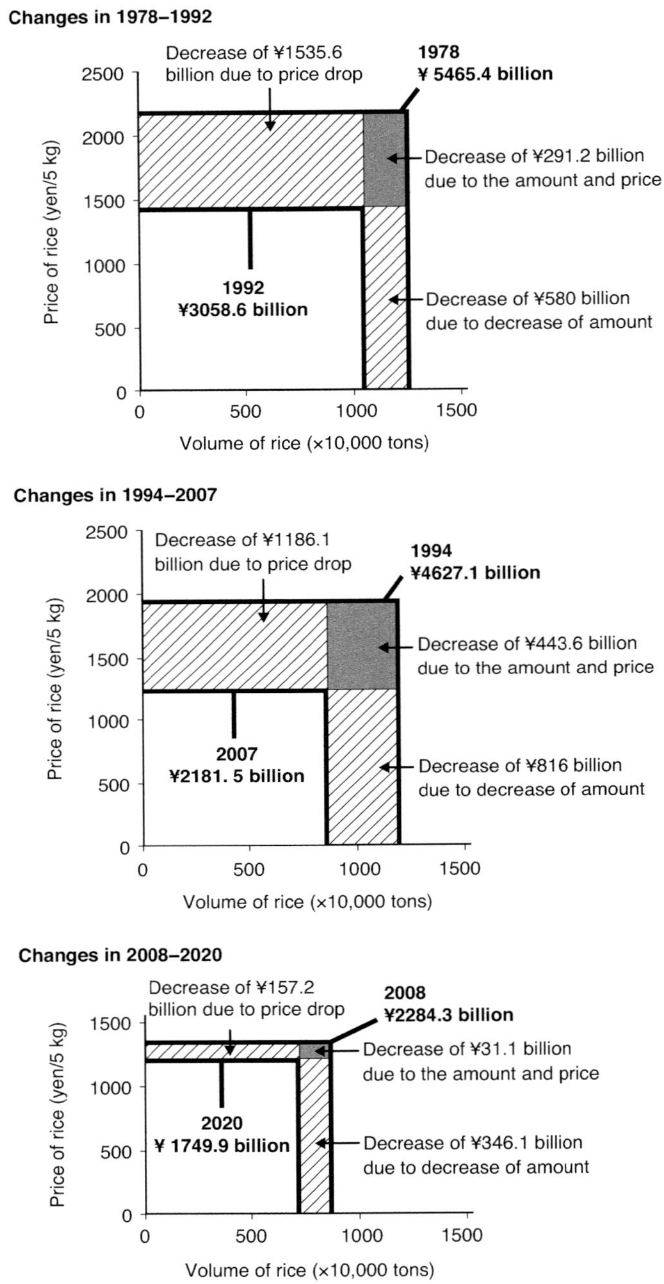

Fig. 6.3. Appraised value of produced rice. *Source*: based on the Ministry of Agriculture, Forestry and Fisheries' *Sakumotsu Tōkei* [*Statistical Survey on Crops*] and *Kome wo Meguru Jōkyō ni Tsuite* [*On Situations Surrounding Rice*]. *Note*: rice price is deflated by the Consumer Price Index of 2020 (excluding owners' equivalent rent). Rice prices in 1978 and 1992 are the price at which the national government bought rice from rice farmers. Prices after 1994 represent the price at which rice was traded through bidding and negotiated transactions.

Survey Department of the Imperial Japanese Army General Staff Office paid special attention to the suitability of terrain for an army marching in wintertime: soldiers, military tanks, and canons could go through dry paddies, soldiers and military tanks could march through wet paddies, and none of them could pass through deep-mud paddies. Before improvement of rice paddies throughout Japan was fully conducted, some of them were waist or even chest-deep in water, necessitating the development of various cultivation methods including, for instance, use of special wooden clogs and boats.

From the viewpoint of agriculture, attention has been paid to whether it is easy to use oxen, horses, or tractors to plow a paddy and whether a paddy can grow wheat in wintertime and upland field crops in summertime. In a statistical survey conducted in 1953, a time of persistent food shortages, the timing for plowing in wheat seeds was judged on the condition of a paddy 3–4 days after normal rainfall. The survey recorded paddies filled with water to above the ankle as wet paddies, and those that were either dry or moist but not showing clear footprints as dry paddies.

For a statistical survey conducted around 1963, when tractors began to be used widely, the standard of judgment was the level of groundwater that would affect the efficiency of a tractor. Wet paddies were those with a groundwater level less than 40 cm from the paddy's surface, semi-wet paddies had a groundwater level of 40–70 cm, and dry paddies were those with a groundwater level deeper than 70 cm.

In 1969, when the policy to curb rice production commenced, yet another new standard was added: whether a paddy could grow non-rice produce in the summertime. The statistical survey at that time classified rice paddies that drained very little or no water in summertime as wet paddies. Paddies for which farmers could control the amount of water drained and the surface of which could be dry in the summertime a few days after drainage were classified as dry paddies.

Today, in accordance with the Basic Survey on Agriculture Infrastructure Information of the Ministry of Agriculture, Forestry and Fisheries, it is general practice to classify paddies from which 4 hours of rain can be drained in 4 hours and for which the groundwater level is 70 cm as dry paddies; paddies with a shallower level of groundwater are classified as wet paddies. The classifications are based on paddy drainage capacity, which is what determines whether a paddy can produce upland field crops.

(Top) Rice planting in a waist-deep wet paddy in Kamiichi-machi, Toyama prefecture (*The Mainichi Shimbun*, June 12, 1954). (Used with permission from Mainichi Shimbun) (Bottom) Restored rice paddy, called horita, a special method for cultivating rice paddies in lowland swamps. Rice planting from aboard a rice paddy boat is demonstrated for local elementary school pupils to experience this traditional rice-growing method. From a collection of Kaizu-shi Folk Culture Museum in Gifu prefecture (*The Asahi Shimbun*, June 3, 2009). (Used with permission from Asahi Shimbun)

Reasons wet paddies exist

Water and soil are the two determinants of a wet paddy. In addition to geographical conditions, in Japan, which is humid nationwide, differences in soil composition greatly affect crops. Soil type and its geographical distribution clearly differ between rice paddies and dry cultivated fields in Japan. In terms of soil condition, 31% of rice paddies in Japan are made of gley soil with poor drainage, 37% are gray lowland soil with inferior drainage, and 9.5% are wet andosol soil resulting from poor drainage. These soil conditions differ significantly from soil in upland cultivated fields, of which 47% are composed of permeable andosol soil.

Shimizu Hiroshi, an agricultural mechanical engineer, points out that improvement of rice paddies triggered the emergence of new wet paddies. Before the Pacific War, a standard rice paddy was 10 ares wide. As the use of larger machinery became systematized for rice growing, the standard paddy size grew wider to reach 30 ares (30 m by 100 m) in 1963. This plot size was proposed on the basis of distance between drainage facilities. Because drainage capacity differs depending on the type of soil, as we've seen, it was necessary to adjust the distance between facilities on a case-by-case basis, making 30 ares for a rice paddy a mere guide. When implementing actual improvements to rice paddies, however, local governments more often than not imposed a uniform paddy size of 30 ares regardless of soil conditions. This resulted in rice paddies with inadequate drainage facilities. Also, regular use of heavy machinery year in and year out has trod down the soil in rice paddies, damaging the drainage in some of them.

Rice paddy history as seen through wet paddies

The characteristics of the four turning points in the Japanese people's relations with rice become much clearer when we focus on wet paddies. Originally, rice paddies had been cultivated in swamps, which explains the prevalence of wet paddies. Prior to the first turning point, however, efforts were made to cultivate dry paddies by such means as underdrainage. This was partly to make it easier to grow upland field crops but, more importantly, dry paddies were more suitable for rice production to begin with.

If a rice paddy is always filled with water, the soil becomes deficient of oxygen. This makes the rice plant liable to disease partly due to the interference of microorganisms. To prevent this, rice farmers prime the soil to take in oxygen by deliberately interspersing a period of no water in the paddy during the rice plant's growth cycle. This practice forces the rice plant's roots to spread out in quest of water, making it easy for the plant to absorb nutrients in the soil and to become more resistant to falling over. The special features of the rice plant's stalk, leaves, and roots means that it can be grown in water-filled paddies that are never drained. But growing rice in a dry paddy where the amount of water in the paddy at any given time can be controlled produces healthier rice.

And that is not all. Converting a wet paddy unsuitable for draft animals into a dry paddy makes it possible to use those animals to plow easily, thus reducing the labor needed to grow rice. Because a farmer can plow deeper furrows in a dry paddy, more fertilizer can be used in the same area, which results in a larger and more stable harvest. Given these merits, it became popular to convert wet paddies to dry paddies during the Meiji period (1868–1912).

Thus, dry paddies are preferable to wet paddies for growing rice. The only reason wet paddies have survived is that rice was in short supply for the Japanese people. Assisted by upgrades to the government subsidy for rice paddy improvement and the widespread use of machinery, the drive to convert wet paddies to dry paddies continued. After the first turning point in 1967, Japan began to experience rice surpluses. After the second turning point in 1978, discussions on the ideal usage for rice paddies grew active on the premise of the merits of dry paddies, including the relative ease of growing upland crops in dry paddies of all the surplus rice paddies and the method to alternate rice and upland crops at regular intervals in the same dry paddy.

The characteristics of wet paddies motivated rice farmers to unite in mutual support. Under the policy to curb rice production, it had been a general practice for local municipalities to assign surplus rice paddy area to individual farmers in equal proportion. A sense of inequality grew between those rice farmers whose area contained more wet paddies compared with those whose area was mostly dry paddies. To cope with this imbalance, new community bonds emerged among rice farmers that saw them contributing to mutual aid funds and promoting concentration of rice farming in regions abundant with wet paddies that can only produce rice, while letting regions full of dry paddies engage in upland cropping.

It should be pointed out, however, that these cooperative efforts among rice farmers did not succeed in using up all the available wet paddy area. Analysis of 22 prefectures by agrologist Takata Yūsuke and his associates reveals that, while rice paddy area as a whole dwindled by 12% from 1972 to 1992, the area of rice paddies with poor drainage decreased proportionally more: those years saw a 21% decrease in wet paddies with gley soil and a 17% decrease in gray lowland soil paddies. Even with the government subsidy for growing rice for processing, it was easier for the owners of wet paddies to convert them into residential sites or leave them abandoned instead of struggling to produce upland dry crops in paddies that can no longer be used to grow rice.

At the third turning point in 1993, even with the heightened interest in the multiple functions of rice paddies, little attention was paid to the difference between wet paddies and dry paddies. Around the fourth turning point in 2008, when the production of upland dry crops in surplus rice paddies faltered, rice for animal feed was introduced as a crop to be grown in rice paddies or in regions unsuitable for wheat and soybeans. Rice for animal feed came on stage when the movement to convert rice paddies to dry paddies reached its limit. Since 2014 to date, rice for animal feed has been grown more to reduce the production of rice for meals.

Troublesome wet paddies

Aside from being less suited to growing upland crops, wet paddies have two more troublesome characteristics.

First, because dry paddies are more suitable for growing rice both in terms of their capacity for nurturing the rice plant and the amount of labor required, rice for meals is bound to be produced more in dry paddies, which means that wet paddies occupy a larger proportion of surplus rice paddies. For this reason, a division of labor between wet paddies, where only rice can be grown, and dry paddies, which can grow both upland crops and non-rice produce, failed to be established.

The second troublesome characteristic is that it is difficult to objectively and quantitatively grasp the problems caused by wet paddies. A range of wet paddies exists: from those that can be made usable with installation of a nullah or a culvert for better drainage to those that are hopeless no matter what effort is made. Additionally, some wet paddies are only troublesome seasonally, like those rice paddies that become muddy only during the rainy season or for a few days after a torrential rain.

How troublesome wet paddies are depends on how they are used. For instance, wet paddies where it is difficult to grow upland crops can be made usable for growing rice and lotus root or for fish farming if the water level is controlled to some extent. Attached to a crawler with a wider contact area, a tractor can be operated in wet paddies without sinking in the mud. When a huge amount of investment and labor are poured in, wet paddies can be made less troublesome.

How a wet paddy becomes a nuisance depends on many factors, such as the condition of the paddy, timing of its use, the effort individual farmers put in, types of crops to be produced, the type of machinery used, investment and capital recovery plans, and labor hours.

One reason it is difficult to judge if wet paddies have been problematic is the influence of the government subsidy for surplus rice paddy area. Surplus rice paddies planted with upland crops are not always dry paddies. Some rice farmers grow upland dry crops in unfavorable wet paddies by relying on the subsidy or in the hope of receiving the subsidy. On the other hand, all the rice paddies in which rice for processing or new-demand rice is grown are not necessarily wet paddies, because those types of rice can also be grown in dry paddies.

Status of rice paddy improvement

For the reasons noted, it is difficult to accurately grasp where wet paddies are found and how they are actually troublesome. Nevertheless, it is possible to calculate a rough estimate of wet paddy area based on data on the status of rice paddy consolidation. Of all rice paddies in Japan today,

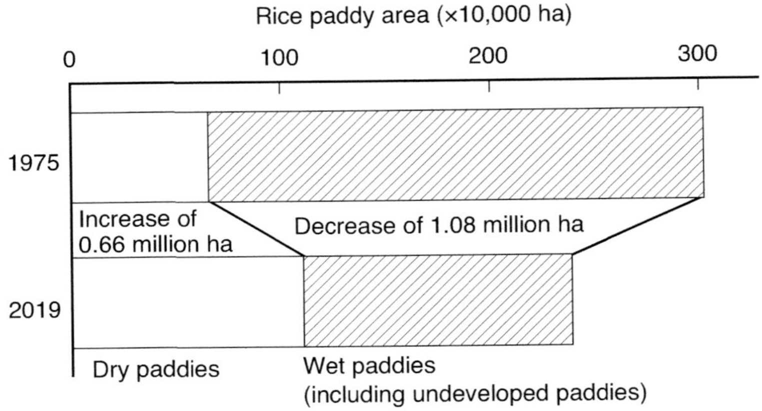

Fig. 6.4. Rice paddy situations in 1975 and 2019. *Source*: based on the ministry of Agriculture, Forestry and Fisheries' *Tochi Riyō Kiban Seibi Kihon Chōsa* [*Basic Survey on Land Use for Infrastructure Development*] and *Nōgyō Kiban Jōhō Kiso Chōsa* [*Basic Survey on Agricultural Infrastructure Information*].

1.59 million ha, or 66% of the total, are plots larger than 30 ares. Dry paddies account for 1.1 million ha, or 70% of these plots. Calculating from the premise that unconsolidated rice paddies remain as wet paddies, then we can estimate that dry paddies occupy 46% of all rice paddies nationwide, with the remaining 54% being wet paddies.

Compared to 1975, or 3 years prior to the second turning point, dry paddy area increased by 0.66 million ha while wet paddy area decreased by 1.08 million ha (see Fig. 6.4). Although this reduction in wet paddy area is huge, many wet paddies still remain even after 44 years (as of 2019) since the launching of the rice paddy usage reorganization measure.

The Japanese people, rice, and rice paddies have a long association. It is not an overstatement to say that Japan's history has been a history of rice production. Taking into consideration the fact that large areas of wet paddies have existed since prehistoric days, we can say that, for the Japanese people, being bound by rice paddies that can only produce rice is also part of their history.

Are wet paddies a burden for Japan's agriculture?

Nowadays, when the link between rice and rice paddies is no longer unquestionable, some scholars regard wet paddies as a burden for Japan's agriculture because of their strain on the national budget. Today, rice paddy area used to grow rice for meals has fallen short of 1.4 million ha, an area nearly equal to the 1.1 million ha of dry paddies consolidated into plots wider than 30 ares. Some people, therefore, point to the cost savings possible if various rice paddy subsidies were eliminated through the bold step of simply abandoning unimproved rice paddies and wet paddies.

Also, considering the burden of managing and maintaining ditches and irrigation facilities, it would be more efficient if only those rice paddies designated for cultivation were grouped together instead of having areas where used and unused paddies are mixed together. For this reason, some scholars suggest that rice paddies to be kept for the future should be carefully selected. One way to do this would be to abandon paddies in regions with many individual wet paddies but keep large-scale rice paddies even though they are wet paddies.

The proposals mentioned all appear quite reasonable from the viewpoint of family finance; that is, to cut expenditures when money is short. From the viewpoint of agricultural management, however, these proposals should be considered in a wider perspective that includes verifying the validity of the logic underpinning the proposed policy. Doing so will clarify both the reality of the rice paddy surplus and wet paddies as well as any problems with current policies.

Problems with Current Policies

Characteristics of policies concerning rice and rice paddies

The policy to curb rice production was not only a policy that converted the rice surplus problem into a rice paddy surplus problem. It was also a policy that promoted the use of surplus rice paddies to achieve food security for Japan. But because the policy was not coercively enforced and because of budgetary constraints, subsidies for surplus rice paddies were suppressed at low levels. While on the surface surplus rice paddies were assessed positively as being eligible for government subsidies, in actuality, only limited areas of surplus rice paddies received subsidies.

The most fundamental characteristic of these policy measures was the incongruence between the target of assessment, which justified the policy measure, on the one hand, and the actual target to be supported by the policy, on the other (see Fig. 6.5). A closer look at this incongruence reveals that wet paddies in particular influenced policy and sheds light on the problems with the current policy.

Rationale of the current policy

The importance of rice paddies is explained as follows in the policy called the Management Income Stabilization Measure:

> In Japan, with its narrow national land and limited cultivated farmland, it is imperative to maximize effective utilization of rice paddies, a highly sustainable production mechanism, in order to secure a stable supply of rice, the nation's staple food, improve food self-sufficiency, and maintain and strengthen the multiple functions of agriculture.

In this statement, both rice paddies used to grow rice for meals and surplus rice paddies designated as such to prevent rice surpluses are equally valued for their roles in food security and the multiple functions of agriculture. The statement emphasizes the importance of using paddies to the maximum degree.

As the legal base to the current policy to curb rice production, the Act on Stabilization of Supply, Demand and Prices of Staple Food aims to stabilize both the price of rice and the balance between the amounts of rice produced and consumed. As such, the law states that the state deems it its own role to support rice farmers who are voluntarily making an effort to prevent rice surpluses and to promote the production of non-rice produce in surplus rice paddies.

For the state, to support these farmers means to help replace the rice surplus problem with a rice paddy surplus. It is not an assessment of rice per se or of rice farmers themselves, but an assessment of rice paddy usage that provides the rationale for the policy. Although the emphasis on a stable supply of rice, the nation's staple food, and its connection with food security may give the impression that rice per se is being positively assessed, in actuality how rice is rated is deeply affected by the weight given to rice paddies.

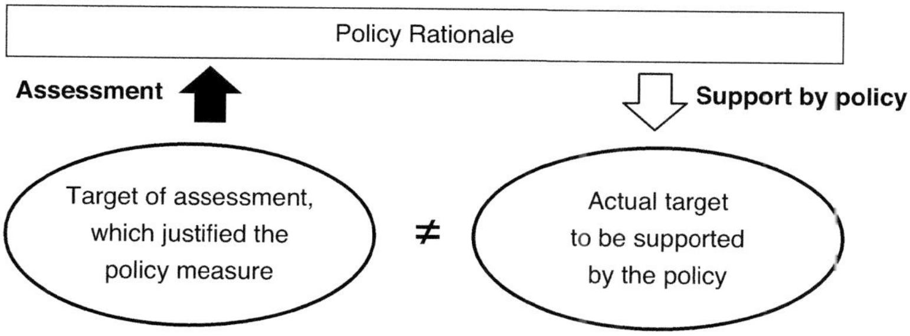

Fig. 6.5. Policy rationale and targets of assessment and support by policy.

Food security and rice for staple food

Rice for meals is technically referred to as 'rice for staple food,' not 'rice to be cooked for meals.' Usage of the term *rice for staple food* stems from the fact that rice is the only cereal produced in Japan's farmlands in amounts that surpass the massive amounts of rice that are actually consumed. It is true that the amount of rice the Japanese people consume is dwindling, but this is only a transient, peacetime phenomenon. The term *rice for staple food* connotes the notion that should food imports be terminated, the Japanese people will produce and eat rice to survive. This idea was used both domestically and internationally to underscore the importance of rice paddies. At the third turning point in 1993, the idea served as justification to treat rice as an exception to international trade rules and to maintain high tariffs on rice imports.

Rice of course still remains the staple food for the Japanese people. And this is not only because the Japanese actually do consume a lot of rice today, but also because it is presumed that rice would be widely eaten in large amounts in times of emergency. In order to make it possible to produce enough rice domestically in a time of emergency, it is imperative that surplus rice paddies be preserved. The rationale behind maintaining a stable supply of rice and acknowledging the importance of rice for Japan's food security is not based solely on the significance of rice. It also includes the value of rice paddies.

Strong relations between wet paddies and Japan's food security

As mentioned, Japan's food security is deeply connected with rice paddies. Wet paddies in particular are related to Japan's food security in two interesting ways. First, food security provides the rationale behind the social significance of protecting wet paddies. Second, it is the prevalence of wet paddies that makes rice paddies of such vital importance to Japan's food security.

Even if rice is not currently being produced in a wet paddy, which can grow nothing but rice, the paddy can still play a role in Japan's food security because it becomes a place to grow rice for meals in times of emergency. I must add, however, that if we disregard nutritional balance and consider calories alone, double cropping with rice and wheat or growing only sweet potatoes can support more people per area of rice paddy than producing only rice. If only all rice paddies were dry paddies, which are amicable to dry upland crops, then the area necessary from the standpoint of food security could be much smaller.

In reality, wheat cannot be produced in those regions in Japan that become snow-covered in wintertime. Also, sweet potatoes and wheat are particularly vulnerable to moisture damage, so they cannot be grown healthily in wet paddies. For Japan as a whole, the more area there is of rice paddies that can only produce rice – such as wet paddies – the greater the area of rice paddies that needs to be protected from the standpoint of Japan's food security.

Under the current policy, Japan's food security is one rationale for stressing the importance of full utilization of rice paddies. Yet the difference between wet paddies and dry paddies in terms of their relation to food security goes unrecognized when considering them as targets of policy support. So, it has to be said that there is a subtle misalignment between the target of assessment that provided the rationale for the policy, on the one hand, and the target of policy support, on the other. While this misalignment seems to be an insignificant matter, the deeper one looks into other issues than food security, the clearer the serious effects that the situation has on Japan's agriculture become.

A hypothesis on the multifunctional roles of agriculture and rural areas

Agriculture's contribution to Japan's gross domestic product is merely 0.8%. Even so, the Ministry of Agriculture, Forestry and Fisheries has not yet been integrated into the Ministry of Economy, Trade and Industry for the two reasons that agriculture produces commodities called agricultural produce and brings a variety of blessings to the Japanese people that money cannot buy. 'Blessings' refers to the multifunctional roles of agriculture and rural areas. These multifunctional roles have provided solid rationales when advocating the raison d'être of domestic agriculture.

We must note, however, that certain aspects of the multifunctional roles of agriculture and rural areas are untouchable. For one thing, agriculture's functions cannot be delineated, because we may not yet be aware of them all. Also, assigning a monetary value to any function is nothing more than just one of many evaluation criteria and, as such, monetary value can be used only as reference. It is difficult to quantitatively assess the multifunctional roles of agriculture and rural areas because some roles, such as their contribution to agrarian culture and biodiversity, don't lend themselves to assessment based on monetary value. Additionally, negotiations at the World Trade Organization have not gone as smoothly as expected and it is not yet confirmed to what degree functions of agriculture and rural areas will be included in non-trade concerns. It will be tricky to deal with the multifunctional roles of agriculture and rural areas as negotiations proceed because they themselves are the non-trade concerns. It would be utterly out of the question to separate out individual functions of each multifunctional role to assign monetary values, or to add up assigned monetary values of individual multifunctional roles.

Although the multifunctional roles of agriculture and rural areas are extremely important as policy rationales, their untouchable aspects obscure clear perception of their substance. Fig. 6.6, prepared by the Ministry of Agriculture, Forestry and Fisheries, confirms this obscurity. Only 65% of rice paddies in Japan produce rice, but this diagram shows that rice is produced in almost all rice paddies. The image of wondrous and beautiful agriculture and rural areas is highly effective in appealing to the Japanese people on the importance and richness of agriculture and agrarian villages. Still, the gap with reality should not be overlooked.

Given the situation, I dare to do the unthinkable and assign monetary values to the multifunctional roles of agriculture and rural areas. I refer to the proforma calculation on the multifunctional roles of agriculture and rural areas published in 1998 by the National Research Institute of Agricultural Economics

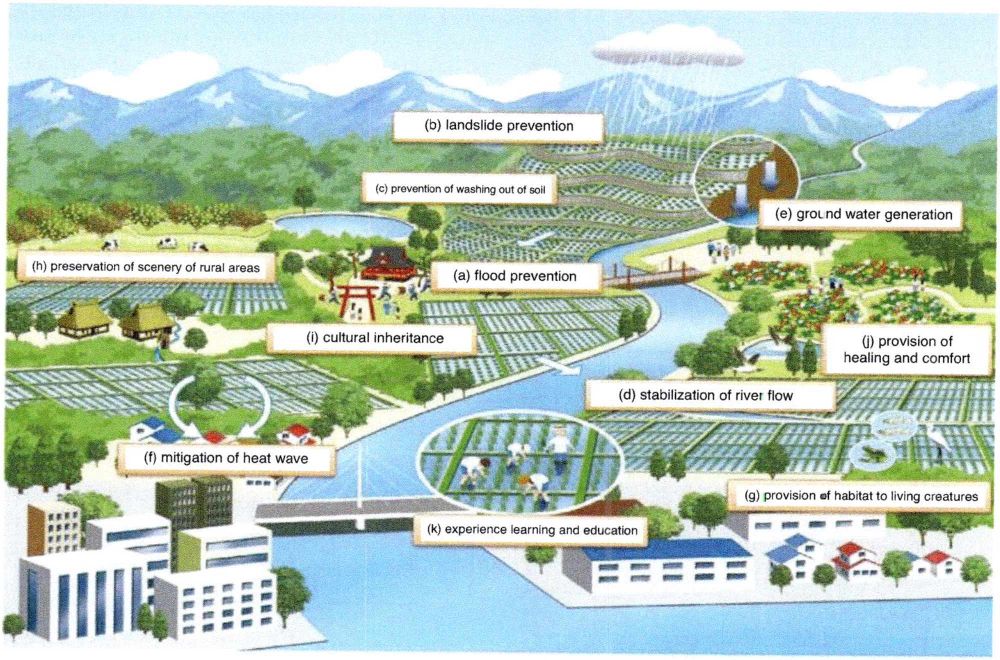

Fig. 6.6. Illustrated image of the multifunctional roles of agriculture and rural areas. *Source*: https://www.maff.go.jp/j/nousin/noukan/nougyo_kinou/.

(the present-day Policy Research Institute of the Ministry of Agriculture, Forestry and Fisheries), the only state policy research institute in agriculture, to attempt to compare the difference in monetary value between rice paddies and cultivated dry fields. Elements not suitable for appraisal, including Japan's food security, agrarian culture, and biodiversity, are excluded. For details, see Ogawa (2017, 2018).

According to my preliminary calculations, at the time of the 1998 report, rice paddies were assessed at an annual monetary value of ¥5292.5 billion, much higher than the ¥1586.3 billion for cultivated dry fields. The per hectare value was ¥1,928,043 for rice paddies and ¥691,826 for cultivated dry fields.

There is a reason rice paddies are evaluated as being about 2.8 times more valuable than dry cultivated fields. A rice paddy surrounded by strong, high ridges can store much more water, thus strengthening its function to prevent flooding and stabilize river flows. Also, a rice paddy filled with water can hold on to soil when it rains as well as mitigate summer heat by evaporation. A sum of ¥4630 billion, or 77% of the total monetary value for rice paddies, was attributed to the presence of ridges. The value based on the fact that rice paddies are filled with water was ¥1417.3 billion. This shows that rice paddies and rice production have contributed greatly to the overall assessment of the important multifunctional roles of agriculture and rural areas.

Incongruence between assessment and support

Next, let us direct our attention to the policy to fully utilize rice paddies, which is increasingly strengthening its air of being synonymous with the policy to curb rice production. This is to confirm the incongruence between the subject being positively assessed for a policy measure and the target of actual support (see Fig. 6.7).

The rationale for the policy stressing the importance of rice and rice paddies is based on appreciation of rice farmers' voluntary efforts to contribute to the stabilization of rice distribution and the rice price by promoting voluntary production adjustment, as well as on the high value attached to the roles that rice paddies and rice production play for food security and realization of the multifunctional roles of agriculture and rural areas.

Yet the support actually offered through the policy is targeted at efforts to curb production of rice for meals and to grow non-rice produce in surplus rice paddies. Also, given its budgetary constraints, the Japanese government has

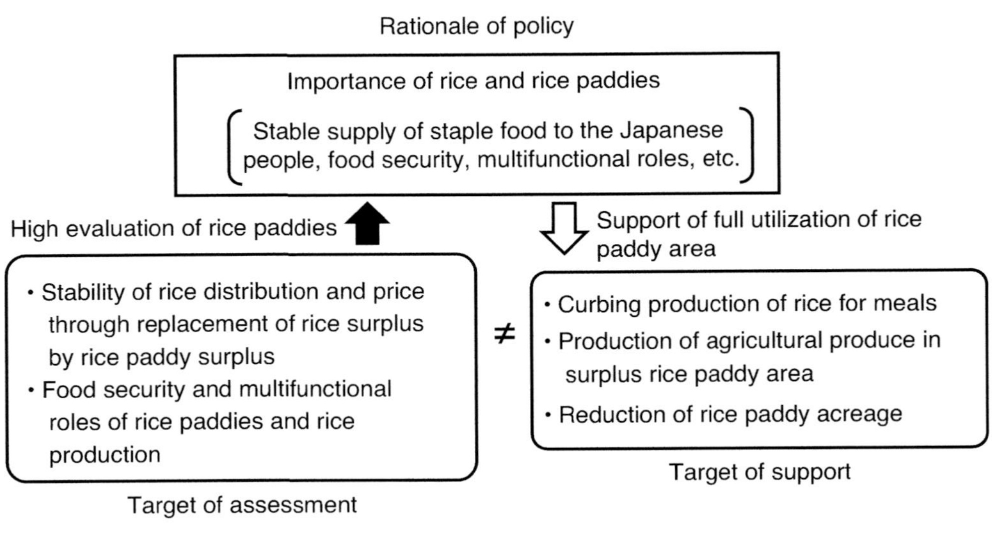

Fig. 6.7. Characteristics of the current rice paddy full utilization policy.

actively pursued the reduction of rice paddy area. So, while the subject of the policy (rice and rice paddies) and the objects of support (rice paddy usage and area reduction) overlap here and there, assessment of the significance of rice paddies has not been sufficiently reflected in choosing actual targets for support. This is true even though appreciation for rice paddies has provided the rationale for a policy stressing the importance of rice and rice paddies.

Merits and demerits of continued positive evaluation of rice paddies and rice production

This incongruence has occurred mainly because the substance of Japan's food security and the multifunctional roles of agriculture have been perceived as an obscure abstract instead of a rigorous standard. For instance, monetary assessments of the multifunctional roles of agriculture and rural areas had been frequently conducted around the third turning point in 1993 and around 1999, when the Basic Act on Food, Agriculture and Rural Areas was enacted. The act strongly advocated realization of the multifunctional roles of agriculture and rural areas as a philosophy. Aside from the study by the Mitsubishi Research Institute introduced in Chapter 4 (see section The Day the Rice Price Drop Became Destined) and the aforementioned estimates by the National Research Institute of Agricultural Economics, the Science Council of Japan also attempted to assess the monetary value of the multifunctional roles of agriculture and rural areas. No full-scale re-evaluation has been conducted since around 1999. This means that the multifunctional roles of agriculture and rural areas have not been re-evaluated every year. As a rationale for policy, then, their importance has passed into obscurity.

The policy based on an obscure rationale for the multifunctional roles of agriculture and rural areas has two characteristics. First, while it is rice paddies and rice production that have contributed greatly to the positive assessment of the multifunctional roles of agriculture and rural areas, it is the assessment itself that brings significant benefit to agriculture and rural areas as a whole in the form of government support. Despite the taboo against separating out the individual substances of the multifunctional roles, being assessed for a policy measure and being the object of actual support do not always coincide.

Second, by passing on the greatness of the multifunctional roles that agriculture and rural areas have played as an obscure rationale, the incongruence between the subject positively assessed as being eligible for a policy measure and the actual target of support becomes larger as time goes on. Even if rice paddy area and rice production continue to decrease, the government can continue to support agriculture and rural areas based on the same policy rationale, paying no heed to the widening gap (see Fig. 6.8).

One can speak about the beauty of the multifunctional roles that agriculture and rural areas have played mainly on the basis of the social evaluation of rice paddies and rice production in the past. Moreover, one can still continue to talk about the beauty of the multifunctional roles of agriculture and rural areas even when rice paddy area and rice production are dwindling.

For instance, the estimates by the National Research Institute of Agricultural Economics were made when rice paddy area overall was 1.13 times larger and area for paddies growing rice was 1.35 times larger than today. The gap between the social value of rice paddies and rice production at the time of evaluation and their actual values today is becoming larger as rice paddies and rice production decrease.

If all rice farmers were still growing rice in paddies as before, support to agriculture and rural areas would automatically become indirect support for rice paddies and rice production. As the number of rice farmers who discontinue growing rice increases, however, policy support may not necessarily benefit all rice farmers.

Characteristics of the direct payment measure for hilly and mountainous areas

There are other policy measures in which the subject positively assessed for a policy measure and the actual target of support do not always coincide. The Direct Payment Measure for Hilly and Mountainous Areas took effect in 2000

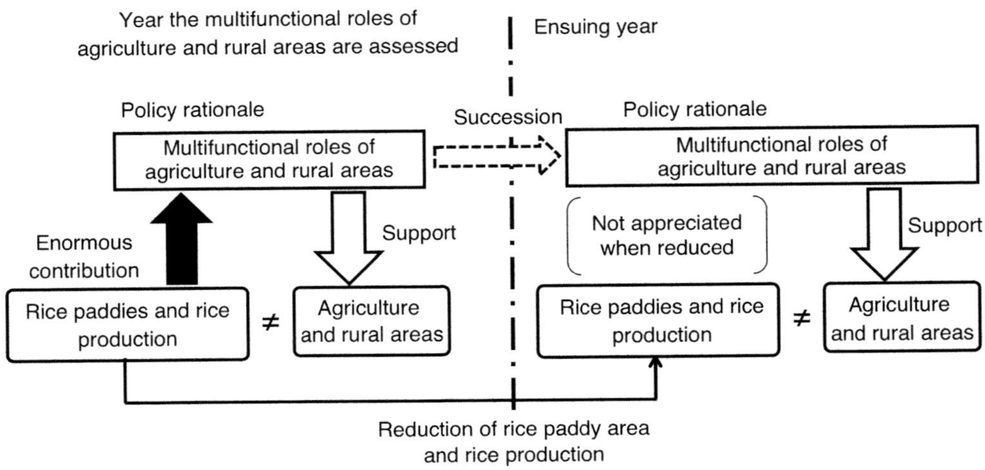

Fig. 6.8. Characteristics of the rationale for the policy on the multifunctional roles of agriculture and rural areas.

with the aim to support the continuation of agriculture in areas with unfavorable conditions to maintain and enhance the multifunctional roles of agriculture and rural areas.

The basic unit value of the subsidy is designed to make up for 80% of the gap in production costs between hilly and mountainous areas and flat areas, albeit subject to seasonal details and conditions. Because the gap in production costs between the two areas is particularly large for rice, greater than for vegetables or pasture, it was decided that rice paddies should receive a higher subsidy than upland cultivated fields or pasture areas. Based on the history of community-to-community connection through rice production, the measure was set out as a policy to support community-wide efforts. The Direct Payment Measure for Hilly and Mountainous Areas was designed on the foundation of uniquely Japanese logic, reflecting the link between rice and rice paddies and the history of rice production.

At the same time, the measure stipulates that the subsidy can be used broadly in accordance with regional conditions. Thus, a region abundant with rice paddies is entitled to receive a larger amount of subsidy but does not have to use the subsidy for rice production. This makes the subsidy favorable to areas full of dry paddies rather than wet paddies because rice farmers there can receive a high subsidy designed for rice production while in actuality growing upland field crops for which the gap in production cost with flat fields is relatively small.

While the basic unit value of the subsidy has been passed on unmodified since the launching of the measure, the gap in production costs between hilly and mountainous areas and flat fields has widened due to technological developments in rice production. For this reason, provision of the subsidy through the Direct Payment Measure for Hilly and Mountainous Areas has failed to fulfill its role of narrowing the gap in rice production costs as before. Although this measure was supposedly designed with full attention to the interrelation between rice and rice paddies, it has proven to be relatively unfavorable to regions where there are many wet paddies.

The relationship between wet paddies and rice for animal feed

In Japan, rice for animal feed has been used in place of corn. Japan has relied heavily on imports for its supply of corn. When both rice and corn can be grown healthily, growing corn is much more productive. Livestock farmers in Japan are much more familiar with feeding corn to their animals. Rice for animal feed is an artificially invented agricultural produce for growing in wet paddies, which are not suitable for upland field crops.

In fact, at the fourth turning point in 2008, production of rice for animal feed was advocated in rice paddies or regions not suitable for upland field crops. But the subsidy for rice for animal feed was disbursed regardless of whether the target of the subsidy was a wet paddy or a dry paddy. As the policy for full utilization of rice paddies increasingly attained an air of being synonymous with the policy to curb rice production, people began to pay more attention to rice for animal feed as a substitute for rice for meals. It became a trend to produce rice for animal feed no matter whether the paddy was wet or dry.

Let us clearly sort out the relationship between wet paddies and rice for animal feed. The fact that surplus rice paddies where growing rice for meals is no longer carried out include wet paddies led to the rationale for the policy that called for rice for animal feed. In actual policy practice, however, production of rice for animal feed has been supported as a means to reduce production of rice for meals, regardless of whether a paddy is dry or wet.

Seeing rice paddies that can produce very tasty rice, like those in Uonuma, Niigata prefecture, and paddies that can produce corn being used to grow rice for animal feed, one wonders if the subsidy for rice for animal feed is not a waste of tax money. This is because rice for animal feed, an invention which was supposedly planted in rice paddies with unfavorable conditions, including wet paddies where it was easy to discontinue growing rice for meals but which were unsuitable for production of upland field crops, has in fact been produced in rice paddies suitable for rice for meals and upland field crops.

Rice paddy or dry cultivated field? That is the question

A rice farmer producing soybeans in rice paddies can receive ¥350,000 more government subsidy per hectare of surplus rice paddy than by growing the same crop in dry cultivated fields. The amount of subsidy is different for growing the same produce in dry cultivated fields or rice paddies. This is because the positive evaluation of rice paddies is reflected in the amount of subsidy.

What, then, is the difference between rice paddies and dry cultivated fields? The difference is whether the farmland can be filled with water or not. This is a realistic difference to demarcate the farmland that can grow rice and that which cannot. At the same time, this difference is compatible with the appreciation that, because of their surrounding ridges, rice paddies contribute more to preventing floods and stabilizing river flow than dry cultivated fields do. For this reason, the practice has been to exclude rice paddies that are apparently devoid of these functions due to damaged irrigation facilities or ditches as targets of the subsidy for surplus rice paddies. This policy is designed to prevent incongruence between the subject positively assessed as being eligible for a policy measure and the actual target of support.

Nevertheless, the presence of rice paddies unequipped with irrigation ditches to start with and totally reliant on rainfall means that the application standard has always been lax. And the range of rice paddies to be subsidized has grown increasingly wider. This allows a farmer, for instance, to temporarily refrain from ridge construction in order to improve productivity for upland crops. Not a small number of rice paddies are without surrounding robust ridges when producing upland crops, have lost ridges due to long-time production of upland crops, or have damaged ditches. Today the subsidy for surplus rice paddies still retains the characteristic of vested interests for rice paddies.

To remedy this looseness of approach, a new condition for receiving the subsidy for surplus rice paddy area was added in 2022 that impacted and confused rice farmers. Now, to be eligible for the subsidy, a rice paddy must have been filled with water at least once during the next 5 years. While it may be justified to subsidize rice paddies producing highly profitable crops, there were a large number of rice paddies that relied on the subsidy for surplus rice paddy area to grow agricultural produce. Rice paddies that alternated growing rice and field crops at regular intervals could continue to benefit from the subsidy. But with this new rule, those rice paddies where alternate growing was not established and, instead, only dry field crops were produced every year ceased to be eligible for the subsidy.

Being capable of being filled with water and actually being filled with water are two very different things. For the sake of simplicity, I will skip giving a detailed explanation. But some surplus rice paddies have been used to grow fruits, flowers, and lawns. Some farmers build greenhouses in paddies or graze their cows there. Such rice paddies may not be surrounded by ridges, and in ordinary times their flood prevention capacity can be sacrificed. In times of emergency, however, it is conceivable that fruit trees would be uprooted, greenhouses taken down, and ridges constructed so that paddies could be filled with water to produce rice. Even so, in normal times with no food crisis, it is hard to imagine a rice farmer interrupting normal activities just to fill a paddy with water in order to be eligible for the government subsidy. To be able to fill a paddy with water in an emergency and to actually fill it with water are worlds apart.

The rule that only those rice paddies that have been filled with water during the next 5 years may be effective in preventing the incongruence between the subject positively assessed as being eligible for a policy measure and the actual target of support when it comes to administrating the subsidies for surplus rice paddy area. But the rule needs to be applied carefully so as not to overlook the broader picture of food security and national land use, for example.

Problems remain even when rice paddy area is reduced

Reducing rice paddy area by converting paddies to dry cultivated fields or mountain forests may appear to be an effective solution to the problem. As rice paddy area dwindles, the government's financial burden in the form of subsidies for surplus rice paddies lessens, making it possible for the government to disburse larger subsidies for the remaining surplus rice paddies. From the viewpoints of national land use for food security and realization of the multifunctional roles of agriculture and rural areas, however, this is not a desirable remedy. It also means that the value of food security for Japan and the multifunctional roles of agriculture and rural areas will go down.

In other words, measures to remedy the rice paddy surplus problem have an inherent dilemma. That is, as a measure becomes easier to carry out as rice paddy area decreases, so its rationale as a measure for surplus rice paddies weakens. And yet, this dilemma has not been talked about much. On the contrary, the reduction of rice paddy area has been actively pursued under the policy of full utilization of rice paddies.

Here is the tricky mechanism behind the connection between the rice surplus problem and measures for handling the rice paddy surplus problem. Apprehension about rice surpluses was mitigated by reducing rice production. But this was by no means a fundamental solution, and apprehension over the surplus in rice paddy area became more serious instead. The problem of surplus rice paddies can certainly be mitigated by reducing paddy area. But this merely substitutes the rice paddy surplus problem with the problems of fragility in food security and an undermining of the multifunctional roles of agriculture and rural areas. Again, this is by no means a fundamental solution. Because the importance of food security and the multifunctional roles of agriculture and rural areas are only obscurely appreciated, however, the issue of surplus rice paddy area has not been perceived as problematic.

Remedial measures to deal with surplus rice paddy area per se and measures addressing the usage of surplus rice paddies from the perspective of food security are essentially two different things. Because these measures have been promoted in an integrated manner under the policy to curb rice production, however, attention has been directed only to preventing rice surpluses in light of the strain on the national budget. Backed by the abstract rationales of food security and the importance of the multifunctional roles of agriculture and rural areas, the policy of full utilization of rice paddies today prioritizes the prevention of rice surpluses, and to a great extent this fact is precipitating the fragility of food security and weakening the multifunctional roles of agriculture and rural areas.

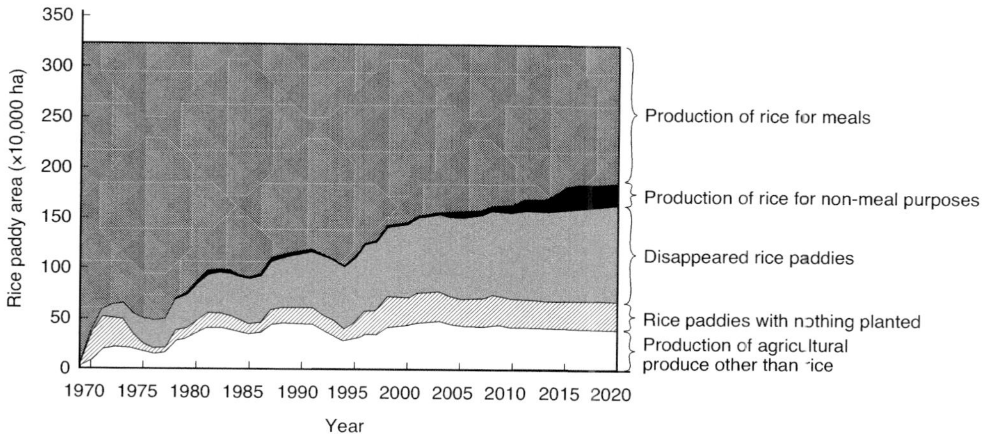

Fig. 6.9. Changes in usage of rice paddy area. *Source*: based on the ministry of agriculture, Forestry, and Fisheries' *Sakumotsu Tōkei* [*Statistical Survey on Crops*] using rice paddy area in 1969 as the standard for the summer situation.

What Detachment of Rice Paddies from Rice Means

Farmland maintained through production of rice

Using 1969, the year rice paddy area peaked, as the base year, let us trace the changes in the usage of rice paddies (see Fig. 6.9). Due to the limited availability of statistical data, I will concentrate on the condition of rice paddies in summertime.

Overall, today's rice paddy area used to grow rice for meals and the area of abandoned paddies are outstandingly large. Together, they comprise 73% of the total rice paddy area in 1969. Here abandoned paddies refers to paddies that have been or could be converted to residential sites, forests, and roads or to those growing crops other than rice.

Area for those rice paddies that made a successful shift from growing rice to growing upland dry crops is only 12% of paddy area in 1969. Rice paddies' departure from rice production has not advanced smoothly even with the drive to promote the production of dry upland crops. The departure from growing rice has been accomplished only through the extinction or disuse of rice paddies.

What, then, happened to those rice paddies that were no longer rice paddies? Let us confirm the trends in rice paddies and dry cultivated fields after 1969 (see Fig. 6.10).

In 1969, there were 3,441,000 ha of rice paddies. Over the years, new rice paddies were added through cultivation, reclamation, and landfill, while old paddies were lost. The old paddies were converted to residential areas, dry cultivated fields, or simply abandoned. By 2016, rice paddy area had dropped to 2,432,000 ha. The most prominent cause of the decrease in paddy area in this period was conversion to residential sites, followed by deterioration and conversion to dry cultivated fields.

The rice paddy area converted to dry cultivated fields was larger than the other way around: 195,756 ha vs 79,369 ha. In 1979, 1 year after the second turning point in 1978, the conversion of dry cultivated fields to rice paddies began to be rigorously restricted. As a result, in subsequent years 129,566 ha of rice paddy area were converted to dry cultivated fields and only 5209 ha of dry cultivated fields were converted to rice paddies. Once the measure to reduce surplus rice paddy area was launched in full, few rice paddies that had been converted to dry cultivated fields were reconverted back to rice paddies.

The greatest reason behind the loss of dry cultivated fields is deterioration. While historically

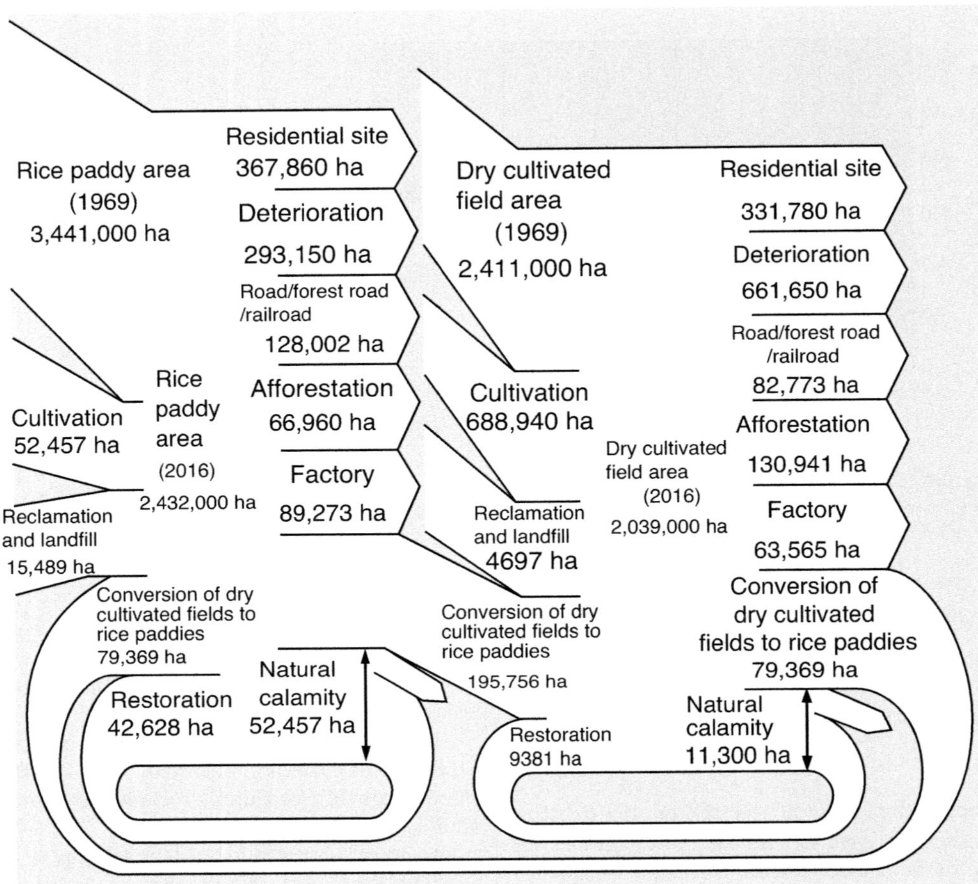

Fig. 6.10. Flow of areas for rice paddies and dry cultivated fields (1969–2016). *Source*: based on the Ministry of Agriculture, Forestry, and Fisheries' *Sakumotsu Tōkei* [*Statistical Survey on Crops*].

farmers, as if to compete with one another, had made their favorable farmland into rice paddies, leaving their relatively unfavorable land to be used as dry cultivated fields, there also have been many cases of farmland that was originally rice paddies being left to deteriorate after conversion to dry cultivated fields.

Also, rice production, which can easily inhibit the growth of weeds because rice paddies are filled with water, is a form of agriculture that does not easily damage farmland. Rice production plays the role of sustaining rice paddies and farmland. In many places, rice has been produced as the last bastion for farmland to be used as farmland.

Cater your stomach to farmland or farmland to your stomach?

The idea of a Japanese-style diet, which was introduced around the time of the second turning point in 1978, was predicated on the notion that one's stomach had to cater to farmland. Yet some argued that farmland should produce food that is strongly favored by locals. This was the notion that farmland should cater to locals' stomachs.

In the half century between the first turning point in 1967 and 2016, how did the Japanese diet and agricultural produce evolve (see Figs 6.11 and 6.12)? There were cases where

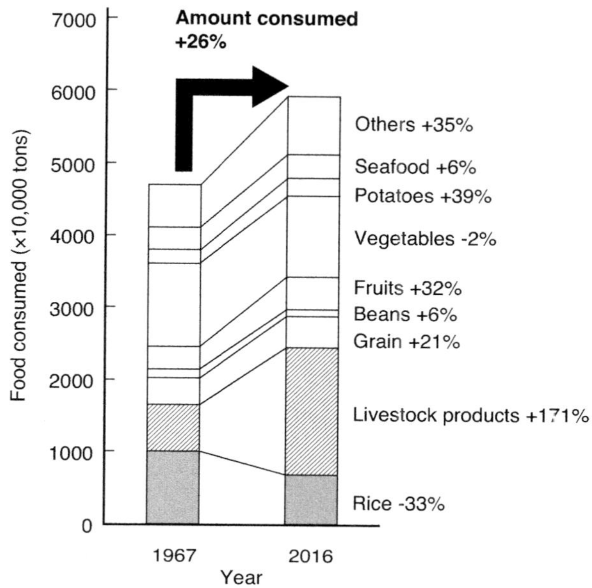

Fig. 6.11. Changes in amount of food consumed by the Japanese people, by item. *Source*: based on the Ministry of Agriculture, Forestry and Fisheries' *Shokuryō Jukyūhyō* [*Food Supply and Demand Balance Sheet*].

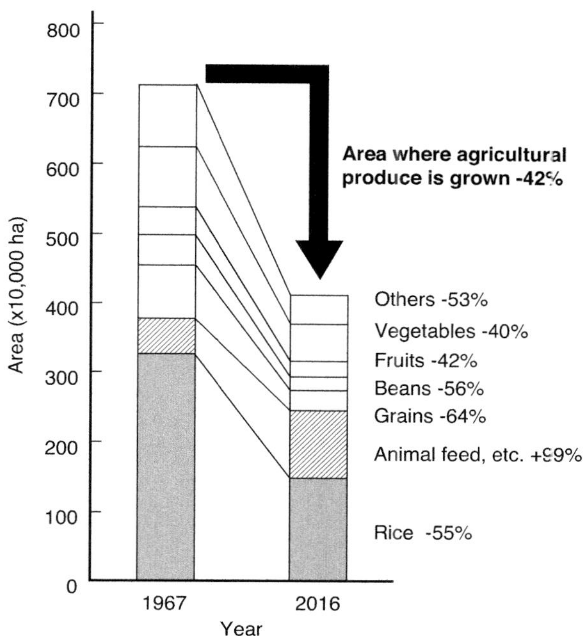

Fig. 6.12. Changes in area where the Japanese people grow agricultural produce, by item. *Source*: based on the Ministry of Agriculture, Forestry and Fisheries' *Shokuryō Jukyūhyō* [*Food Supply and Demand Balance Sheet*].

changes in the usage of farmland and dietary patterns synchronized, such as the decrease of rice consumption occurring in tandem with the increase of meat production and consumption and animal feed production. By far the greatest change, though, was the big drop in area for agricultural production from 7.11 million ha to 4.10 million ha, which occurred as the amount of food consumption per se increased. In this case, there was no sign of synchronization between the two.

Concentrating attention only on rice and rice paddies out of all the properties of Japanese agriculture may mislead one to believe that Japan can make up for the fact that people do not eat as much rice as before by growing other agricultural produce. Some may perceive wrongly that the production of vegetables has not grown because farmers continue to produce rice. In actuality, however, it is the production of agricultural produce as a whole that is dwindling.

Dry cultivated fields like rice paddies and vice versa

In evaluating the degree to which the multifunctional roles of agriculture and rural areas have been realized and whether rice production is viable or not, totally different appraisals can be made regarding rice paddies and dry cultivated fields. It has to be pointed out, however, that it is extremely difficult to accurately judge the classification of land category – whether an area is rice paddies or dry cultivated fields – at a glance even for an expert. Taking the history of a particular farmland's development as background, dry cultivated fields are sometimes treated institutionally as rice paddies. On the other hand, some rice paddies are excluded from the list of targets of the subsidies for surplus rice paddy area. The classification of land category, appearance of the farmland, and institutional treatment of the land do not always coincide. Even some barren rice paddies are still regarded as rice paddies because they are the end product of compliance with the policy to curb rice production or in accordance with some particularly local rule.

(Top) Rice harvesting in a rice paddy (in the center) surrounded by weeds. Even though the foreground as well as areas left and right of the rice paddy are completely covered by weeds, the plot is still registered as a rice paddy according to land category classifications. (Middle) Soybeans grown in a rice paddy. (Bottom) Although the administrative classification of the farmland in this photo is "field" not "paddy field," it is actually watered and used as a paddy field. (Photographer: Masayuki Ogawa)

(Top) Peach trees planted in a rice paddy. (Bottom) A vinyl greenhouse built on a rice paddy. (Photographer: Masayuki Ogawa)

Ruining rice terraces

'It must have been a splendid view had it not been for a fallow period,' and 'I wish I had been here two or 3 years earlier.' It is not unusual to find these kinds of comments on the internet posted by visitors to rice terraces who arrived all excited in anticipation of seeing great scenery. Personally, I am such a fan of rice terraces that I visit various locations to take photos. I must admit that it has become increasingly difficult to take a well-composed photo of a rice terrace. Rice terrace maintenance has already reached its limit. Rice terrace deterioration is no longer stoppable through conventional means such as introducing beautiful views or provoking rice farmers' pride in and the Japanese people's appreciation for rice terraces through such campaigns as the 'One hundred best rice terraces in Japan' or 'Our rice terrace heritage must be passed on.'

If a rice terrace is to be preserved like a museum piece, it would be sufficient to keep some popular rice terraces with splendid views and which are conveniently located in terms of access from urban centers. This, however, will not protect and preserve rice terraces as part of the local culture in localities throughout Japan. When one visits a rice terrace, one's attention may be drawn to the rice paddy where rice is being grown. But it is the deteriorated rice terrace – which got that way because of consumers' dwindling interest in rice – that is the symbol of the unsolved problem of surplus rice paddy area.

The rice terrace is a battlefield between beauty and deterioration. (Photographer: Masayuki Ogawa)

Trends in rice paddy and farmland that ceased to be followed

While before, rice paddies used to be lost to regional development, the population decreases of recent years have squashed development pressure to the extent that in 2015 there were more ruined rice paddies than newly cultivated rice paddies.

Over time, the pattern of decrease in rice paddy and farmland area changed. In 2017, the Ministry of Agriculture, Forestry and Fisheries discontinued surveying reasons for reductions in farmlands in detail, except areas in ruin. Also, the 2020 Census of Agriculture and Forestry in Japan did not study the status of rice paddy usage in detail, and the status of deserted cultivated land, which I am sure readers must have heard of, was not surveyed at all.

The overall trends in farmland, including the aftermath of the surplus of rice paddy area, the usage of rice paddies and dry cultivated fields, and deserted cultivated land, have over the past 5 years or so rapidly grown more difficult even for experts to grasp. The difficulty in grasping the current situation, however, is not an excuse to remain indifferent to rice paddies and farmland. On the contrary, in order to consider what to do about the surplus of rice paddy area, which is the essence of the rice problem in Japan, stronger and closer attention should be paid to rice paddies and farmland.

Limit to Japan's rice policy

The rice surplus problem that the Japanese people faced at the time of the first turning point in 1967 has been replaced by the problem of surplus rice paddy area. Much of the rice paddy surplus problem has already been solved by the disappearance of rice paddies themselves, only to be de facto replaced by the problem of the fragility of food security and the multifunctional roles of agriculture and rural areas. But this problem of fragility has not openly surfaced. This is because the role of food security along with the multifunctional roles of agriculture and rural areas have been upheld only as vague policy rationales. The major shortcoming of Japan's agricultural policy here is that, while rice paddies and rice production in the past contributed greatly to positive evaluation of the state of food security in Japan and the multifunctional roles of agriculture and rural areas, the reduction of rice paddy area and decline in rice production do not lead to an argument for the fragility of food security and multifunctional roles of agriculture and rural areas as policy rationales.

At this point, I could draw a simplistic conclusion that subsidies for surplus rice paddy area should be increased for food security and to highlight the multifunctional roles of agriculture and rural areas. But in actuality there have been many cases of rice paddies falling into ruin after being converted to dry cultivated fields or deteriorating or being converted to dry cultivated fields while retaining the status of rice paddy in land category classifications.

To repeat, the essence of the rice problem in Japan cannot be fully understood unless attention is paid not only to rice but also to the rice paddy problem. And to discuss the rice paddy issue, we must pay attention to farmland as a whole, including dry cultivated fields and national land use.

The surplus of rice paddy area is a serious problem that has been passed on unsolved until today. When and if a next substantial turning point comes about in the future, it should become a turning point that encompasses not only a focused outlook on what to do about surplus rice paddy area alone, but also a wider perspective covering farmland and national land use as a whole.

Next, Chapter 7 will explain the greatest crisis soon to confront the Japanese people – who, in fact, may already have one foot in it – by focusing on the naïve dogma that says Japan's agriculture is necessary for Japan's food security. Chapter 7 will also describe the new turning point for which Japan should be resolutely prepared, and the future beyond.

References

Ogawa, M. (2017) *Suitō no Shiryō Riyō no Tenkai Kōzō* [Social Structure and Development of Using Paddy Rice as Animal Feed]. Nihon Hyōronsha, Tokyo.

Ogawa, M. (2018) Jinkōgensyō shakai ni okeru suiden furu katsuyō no tenbō. Prospects of rice paddy full utilization in a society of dwindling population, *Nōgyō Nōson Kōgakukai-Shi* 86(12), 1111–1115.

7 The Arrival of an Age of Farmland Surplus: The Fifth Turning Point in 2052

Abstract

Chapter 7 predicts the arrival of a fifth turning point. The essence of Japan's rice problem cannot be fully grasped unless the problem of rice paddies is also taken into consideration. Similarly, all farmland, including dry fields, and the national land use must also be taken into consideration when dealing with the rice paddy problem. The analysis in this chapter pays special attention to a rather naïve doctrine that agriculture is indispensable in Japan for food security. The chapter foretells that an excess of farmland will occur by 2052 if the existing farmland continues to be preserved and Japan's population keeps dwindling. Pointing out that the day will come when, from the viewpoint of food security, Japan will experience an abundance of farmland, the chapter urges discussions about the mode of existence for agriculture and national land use in Japan to cope with this future predicament.

Newly Emerging Realm X

Farmland deficiency

Agriculture provides three major elements of food security: farmland, farmers, and agricultural technology. A report submitted to a study group organized by the director of the Economic Security Policy Division, Ministry of Foreign Affairs, makes the following reference to farmland as a composition of food security:

> The most important factor to secure the minimum level of food security is farmland. Although Japan had more than 5 million hectares of farmland before the Pacific War, 70 million Japanese people suffered from starvation. It would be difficult for the farmland in Japan today (approximately 4.61 million hectares) to supply 2020 kilocalories per capita, which is the minimum daily requirement, when a poor harvest is caused by an emergency or when fertilizer imports from abroad are cut off. (Shokuryō Anzen Hoshō ni Kansuru Kenkyūkai, 2010)

From the viewpoint of food security, Japan today is deficient in farmland. Moreover, it takes time and money to restore farmland once it has deteriorated. Even if farmland is successfully restored, it will take time to resume farming when there is no farmer to tend to the land, farming technology knowledge has not been passed on, or agricultural machinery is not immediately available. It is, therefore, extremely important from the viewpoint of food security that all farmlands throughout Japan are continuously engaged in farming.

Introduction of a potential indicator for food self-sufficiency

In March 2015, a new indicator called the food self-sufficiency potential was introduced. Because possible scenarios in times of emergency vary so widely, the indicator temporarily sets aside cases of emergency and calculates the potential to produce food domestically on the premises that labor and production materials are both secured. The indicator was proposed with the aim of sparking momentum for the Japanese people to focus more deeply on the concerns and discussions about food security.

According to this indicator, the current level of the Japanese diet centered around rice, wheat, and soybeans cannot be sustained even if all the farmlands in Japan are fully used. Farmland area would be insufficient even when potatoes, which are high in calories, are the main crop.[1] It is simply impossible for the current area of farmland to produce enough food for the survival of every Japanese.

DOI: 10.1079/9781800627093.0007

Two faces of the food self-sufficiency potential indicator

The food self-sufficiency potential indicator is supposed to provide rationale for the protection of all farmlands. A Japanese daily newspaper reported: "In order to realize the vision of full utilization of barren farmlands, the Ministry of Agriculture, Forestry and Fisheries introduced a new indicator of self-sufficiency potential to give momentum to the restoration of farmland' (Nihon Keizai Shimbun, 2015).

The notion of food self-sufficiency potential, however, has another face. As Japan's population dwindles in the years to come, the notion will transform at a certain point to a rationale for reducing farmland area. As a matter of fact, around the time the agriculture ministry was studying the food self-sufficiency potential indicator, the Ministry of Finance, in the Proposition on the Fiscal Year 2015 Budget Formulation by its Fiscal System Council, expressed a negative view about the continued protection of farmland, which the ministry deemed unnecessary from the viewpoint of food self-sufficiency.

Rising food self-sufficiency potential as population dwindles

Let us consider the relationship between population decreases and farmland's potential to produce food following the line of thinking that gave rise to the idea of food self-sufficiency potential. Here, for the sake of simplicity, a farmland's potential to produce calories and the calories necessary for a person's daily life is used to measure the balance between the population and farmland, assuming that any farmland can produce the same amount of calories as a rice paddy growing rice.

The potential for farmland to produce calories per capita increases as the Japanese population, the denominator of the calculation, decreases (see Fig. 7.1). Even when farmland continues to dwindle, the potential will supersede the per-capita daily calorie requirement in the future.

Arrival of new turning point P and new realm X

In terms of the relationship between population and farmland, the current farmland area falls below the area required to produce calories to support the entire population of Japan, given the calorie production potential per area. As Fig. 7.2 shows, however, this balance will reverse in the future. If Japan retains all the farmland area recorded in 2019, the reversal will take place in 2051 or 2052.

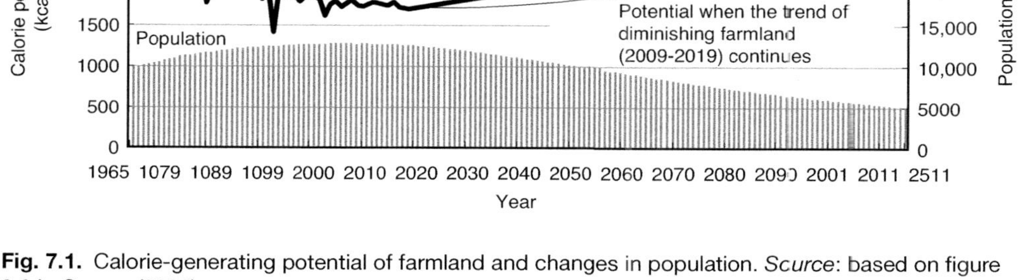

Fig. 7.1. Calorie-generating potential of farmland and changes in population. *Scurce*: based on figure 3.6 in Ogawa (2022).

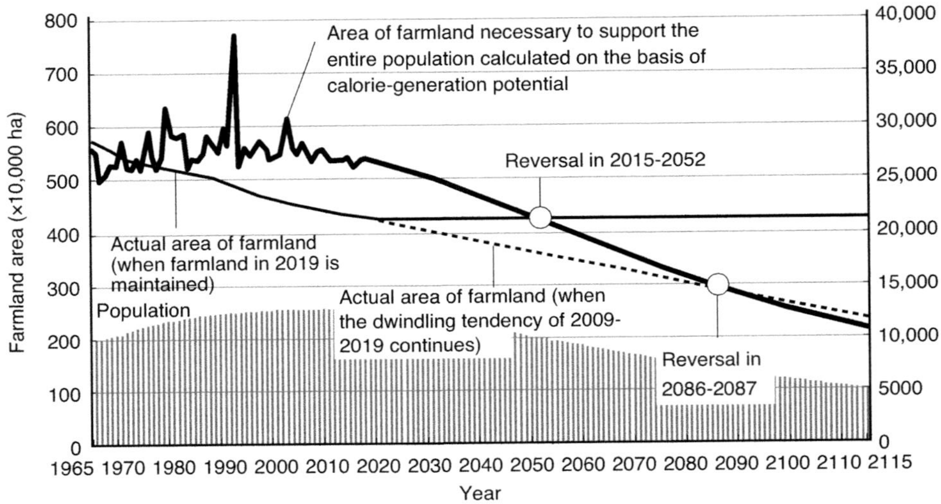

Fig. 7.2. Changes in population and farmland area. *Source*: based on figure 3.7 in Ogawa (2022).

If the farmland needed to support the entire population from the viewpoint of the potential to produce calories can be called 'the minimum farmland necessary for food security,' this minimum area will be smaller than the actual area of farmland expected to exist in 2052, 30 years from the time of writing. Even if the trend of dwindling actual farmland continues, this reversal will take place in 2086 or 2087.

I call the point at which the balance between the minimum area of farmland necessary for food security and the actual area of farmland reverses 'turning point P.' Let me explain with the help of a model (see Fig. 7.3). When the minimum area of farmland necessary for food security decreases along with a population, while the actual area of farmland is maintained, the importance of farmland from the viewpoint of food security will decline over time. From the viewpoint of food security, surplus farmland area will emerge after turning point P. I call this surplus farmland 'realm X.' Realm X will increase year by year along with population decreases.

Referring back to Fig. 7.2, we see that Japan will reach turning point P and face realm X in the not-too-distant future of 2052. The timing of turning point P's arrival fluctuates depending on premises and calculation methods. What is important here is not to accurately predict the specific timing, but to be aware of the possibility that the balance between the minimum area of farmland necessary for food security and the actual area of farmland could reverse sooner or later. Here I refer to one possible timing for turning point P's arrival – 2052 – as the fifth turning point.

Arrival of turning point P delayed as farmland dwindles

Needless to say, if the actual area of farmland dwindles more rapidly than the minimum farmland area necessary for food security, turning point P will never come about and there will be no surplus farmland. The more the actual area decreases, the further away in the future turning point P will become, thus further delaying the arrival of turning point P and realm X (see Fig. 7.4).

For this reason, at this point some readers may think that it is more important to protect the existing farmland, which is in short supply from the viewpoint of food security, than to worry about turning point P or realm X, which may or may not come in the future. This thinking, however, has a pitfall. The luxury of worrying about dwindling farmland when one's

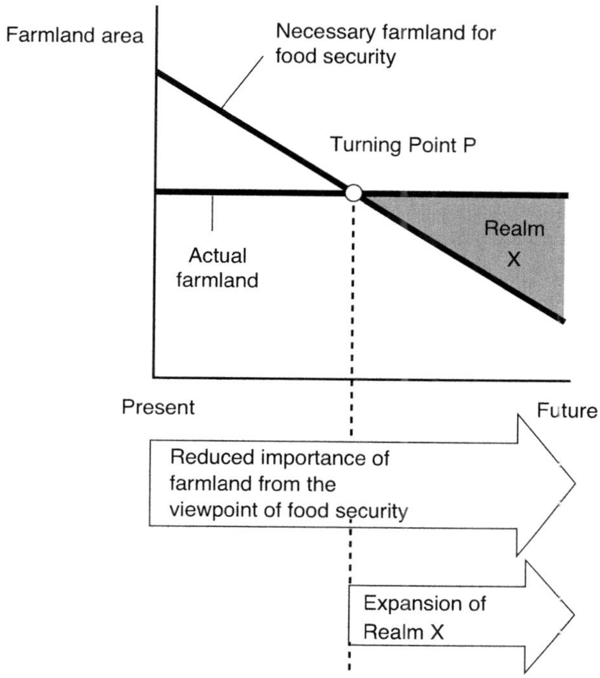

Fig. 7.3. Turning point P occurring as population decreases, and the emergence of realm X. *Source*: based on figure 3.2 in Ogawa (2022).

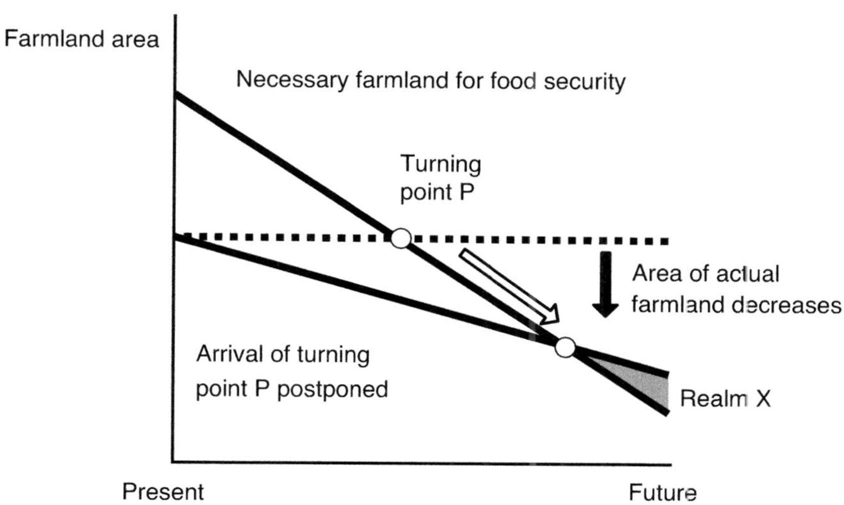

Fig. 7.4. Turning point P moves further away into the future as farmland decreases.

thinking is bogged down in food security issues is only possible when the population continues to grow or when farmland is in short supply as it is today. To advocate for the protection of agriculture and farmland only when the minimum farmland area necessary for food security decreases due to drops in population cannot be a fundamental remedy for the issues at hand – that is, decaying agriculture and deteriorating farmland – unless prospects for the future beyond simply securing enough farmland for food security are prepared.

For those who believe protection of farmland and improvement of agricultural technologies are all for the sake of food security, postponement or non-arrival of turning point P due to rapid decreases in farmland may allow this continued line of thought. Thus, turning point P highlights this flawed thinking and symbolizes the arrival of an era that requires a new perspective.

Under the Basic Act on Food, Agriculture and Rural Areas, a basic plan has been carried out to connect the act with concrete policies. This plan mostly has a target of 10 years and is reviewed every 5 years. The fifth turning point coming in 2052, however, does not fall within the range of this scheme. Moreover, since turning point P may move further into the future, like a road mirage, as farmland continues to dwindle, this basic plan will never be able to deal with turning point P or realm X.

In order to prevent disorderly deterioration of farmland, a new subsidy to support forestation of farmland was introduced in 2022 for the first time as a governmental agricultural policy. To reduce farmland by converting it to forest only delays the arrival of turning point P. It is not a scheme originating from deliberations on what to do about realm X.

The first turning point vs the fifth turning point

What we have discussed so far will be more easily understood when it is compared with the first turning point (see Table 7.1). Japan achieved self-sufficiency in rice at the first turning point in 1967. But because at that time the country did not have a vision on the use of surplus rice paddy area, the policy to curb rice production was launched. That policy included as an emergency measure discouragement of growing anything in surplus rice paddy area. The framework of the policy to curb rice production was molded at the second turning point in 1978 to promote production of non-rice agricultural produce in surplus rice paddies. Half a century since its launching, this policy still has not gotten into gear in many regions, and the problem of surplus rice paddy area is still far from being solved.

With the imminent arrival of the fifth turning point in 2052, I must point out that this time Japan's entire farmland will fall into the same rut as rice paddies did earlier if the usage of realm X is not thought through.

According to future population estimates based on a medium fertility rate and a medium mortality rate, as published by the National Institute of Population and Social Security Research in 2017, 2052 will be the last year in which the Japanese population exceeds 100 million. This is a stark contrast to the first turning point in 1967 when the Japanese population exceeded 100 million for the first time. Both turning points – 1967 and 2052 – share the common issue of what to do with spaces newly deemed as being in surplus.

Table 7.1 shows that the fifth turning point in 2052 will happen 44 years after 2008, the year the Japanese population peaked, thus hinting at the arrival of the next turning point. Coincidentally, this is the same interval between the year when per-capita consumption of rice peaked out in 1923 and the first turning point in 1967. This shows that Japan has ample time to prepare for the fifth turning point. So as not to make the same mistake with the problem of surplus farmland that Japan did with rice paddies earlier, the government this time should do everything to prepare itself for the coming of the new turning point.

A vision for realm X is needed before the arrival of turning point P

What I mean by the heading to this subsection is that the government should draw up a vision for the future before a surplus of farmland becomes

Table 7.1. Comparison between the first and fifth turning points.

	First turning point	Fifth turning point
Timing of arrival	1967	2052
Population	First year population exceeded 100 million	Last year population is over 100 million
Pre-arrival phenomenon	Rice eaten per capita peaked in 1923	Population peaked in 2008
Time lapsed between pre-arrival phenomenon and turning point	44 years	44 years
Newly emerged phenomenon	Rice enters surplus when produced in all paddies	Farmland enters surplus from the viewpoint of food security
Necessary vision	Vision about surplus rice paddy area	Vision about surplus farmland area (realm X)

a reality, not after. Let me explain by using the case of rice for export as an example.

Although international agricultural policy rules stipulate a reduction of export subsidies, the Japanese government today continues to encourage production of rice for export in surplus rice paddy area, under the guise of the catchphrase 'rice to exploit new markets.' So far, there has been no complaint from overseas, allowing the government to continue its de facto export subsidy.

The subsidy for rice for export was launched in full swing in 2018. Ideally, though, at the very latest an earnest policy should have been envisioned prior to the first turning point in 1967, or more than half a century earlier. Had there been a vision for future surplus rice exports while rice was still in short supply, Japan's history of expansion of rice paddy area and rice production would have taken on a completely different meaning. Even if the rice-exporting industry had not developed immediately, half a century would have been enough time to gain experience in marketing Japanese rice, growing rice that catered to overseas markets, and implementing an export method that would not provoke criticism from the United States and other countries in the world.

In actuality, however, partly owing to uncertainty that an era of rice self-sufficiency would come, the Japanese people continued to devote their energy to rice paddy cultivation and rice production in order to settle the immediate problem of rice shortages. And when the first turning point arrived in 1967, people were bewildered in the face of the problems of the rice and rice paddy area surpluses.

From the viewpoint of food security, there is no guarantee that an era of farmland surplus will definitely come. Numerous possible occurrences could affect its arrival: a rising birth rate, increasing immigrants to Japan, aggravated deterioration of farmland, eruption of a war, and so on. In recent years, anxiety about a food crisis has been heightened by the massive volcanic eruption in Tonga in 2022 and Russia's military invasion of Ukraine. Thinking of all the possible occurrences, it is very easy to believe that an era of surplus farmland will never come.

As priority is given to attending to the current shortage of farmland and the accompanying sense of anxiety, discussion on realm X is bound to be of secondary importance. This being so makes holding on to a strong determination to envision realm X prior to the arrival of turning point P all the more necessary. If, instead, the Japanese people continue to protect farmland with no particular sense of purpose and try to increase per-hectare production of agricultural produce solely out of concern over the immediate situation, they will inevitably feel helpless in the face of a future needing to deal with realm X's vast expansion.

Turning point P can be moved

Turning point P shifts in accordance with population, farmland area, and agricultural technology. When advances in agricultural technology bring higher yields from the same farmlands, for instance, the minimum farmland area necessary for food security decreases and turning point P comes nearer (see Fig. 7.5). In other words, by picturing the future prospects that can lead to a ground design for a new form of agriculture and national land use in preparation for the imminent arrival of realm X, Japan can expedite the arrival of turning point P or enlarge realm X, taking advantage of changes in population, farmland, and agricultural technology. Conversely, maintenance of farmland or development of agricultural technology in the absence of a clear idea of the prospects for realm X could unnecessarily cause an increase of surplus or barren farmland.

Japanese agriculture supported by deficiency and anxiety

Previously, there were no terms equivalent to turning point P or realm X. In my opinion, this is because, when evaluating agriculture, the Japanese people attach too much importance to the criterion of affluence brought about by fulfillment and, conversely, the criterion of affluence derived from overabundance has been missing.

Suffering related to food shortages has been constant throughout Japan's history. The threat of famine and the need for families to reduce the number of mouths to feed through, for example, giving up children for adoption or entering them into apprenticeships was ever present. Even after the Pacific War, Japanese relied on imported food to fill their stomachs. That Japan has constantly not been able to feed its citizens with its own agricultural produce has, ironically, nurtured a kind of sense of security among Japanese. That is to say, people's sense of security comes from two things: first, the realization that even if domestic agriculture alone cannot feed the nation, in peacetime day-to-day life has not been affected at all and, second, the assurance that their belief in the importance of domestic agriculture as unquestionable is well warranted because Japan will need to rely on agriculture when food shortages occur in times of emergency.

For instance, some believe that it is not necessary to worry about a low rate of food self-sufficiency because it does not affect people's day-to-day lives. On the other end of the spectrum, as long as a feeling of being impoverished caused by deficiency persists among people, those who are urged to stress the importance of domestic agriculture can feel assured that they can continue their argument. While those two positions may differ in their attitudes toward a low rate of food self-sufficiency, each enjoys their

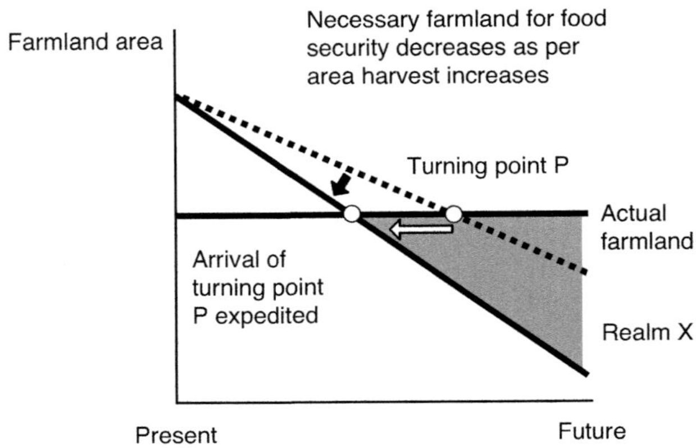

Fig. 7.5. Arrival of turning point P is expedited when per-area harvest increases.

respective sense of comfort generated through their own evaluation of affluence stemming from having enough.

Looked at from a different perspective, all of this means that concern about there being a surplus of food remains low. Even if some people are concerned about it, the issue is not seen as a priority. Any urgency over surplus food is secondary to concerns about deficiency, because there being too much food is thought of as something that can be dealt with when it actually happens.

Barren farmlands must be an incredibly enviable scene from the viewpoint of an era of food shortage. Still, Japanese people today do not get a sense of affluence from this scene. Not having the criterion of affluence derived from overabundance, people do not know what to do with surplus rice paddy area and farmlands.

Ever since the Japanese people accomplished their long-sought-after goal of self-sufficiency in rice, they have had to deal with surpluses of rice and rice paddy area. Just as before, will they sit by idly regarding the imminent problem of surplus farmland area, or will they forge a new path by taking advantage of past experiences and lessons? This is not an issue to consider after turning point P arrives. It is an issue confronting the Japanese now.

Will the Japanese continue to indulge in the strange sense of comfort stemming from the familiar situation of domestic agriculture failing to meet food demand? Will they continue to endlessly shout the importance of agriculture for food security in the midst of dwindling farmland area? Or will the Japanese forge a new path out of the resolve to find new values in surplus farmland? Japan is at an important crossroads.

Solution to the rice problem and an imminent peril

Once Japan has a clear idea of the prospects for realm X, a path to realize the desirable form of farmland and rice paddy area based on that likelihood will present itself. This will lead to settlement of the unsolved problem of the rice paddy area surplus, and, eventually, the rice problem.

The rice problem in Japan will not be solved only by finding a solution to the problem of rice as an actual physical substance or to the problem of surplus rice paddy area as passed on from the past. Japan's rice problem cannot be solved only by attending to present issues with an eye on the past. It can only be solved by tackling the future problem of what to do with realm X.

But there is a deep-rooted problem here: the easiest stance is to postpone the arrival of turning point P instead of squarely facing the problem of realm X. If farmland area continues to dwindle, thus postponing the arrival of turning point P, the Japanese people can continue to indulge in a sense of comfort with the status quo knowing domestic agriculture cannot fill the demand for food.

Current policy has promoted reductions in rice paddy area while at the same time stressing the importance of the existence of surplus paddy area. Similarly, the government encourages conversion of farmland to forest while preaching the importance of food security. These measures appear to be contradictory, but they share a common characteristic. All of them are effective measures to postpone the arrival of turning point P and prolong as long as possible the situation of domestic agriculture being deficient in meeting domestic food demand.

It will be tough to squarely face realm X. This is because appreciating the sense of affluence derived from overabundance calls for deeper communication among people since it is not a norm necessarily shared by everyone naturally, in contrast to the sense of affluence brought about by fulfillment, which is easier for people to empathize with because it synchronizes with human physiology and instinct. To deal with something that defies easy quantification, such as the ability to appreciate affluence, and the determination to accept it, have their own difficulties.

The greatest impending crisis for Japan is the imminent arrival of realm X. It is not some future natural phenomenon that will happen no matter what we do or do not do. Rather, it is a crisis that is already confronting the Japanese people and demanding to know whether they can internalize the sense of affluence derived from overabundance. The unsolved problem of surplus rice paddies is nothing but a canary in a coal mine heralding this greatest crisis.

The Way to Welcoming Turning Point P

Realm X as the logical ground for agriculture and national land use

A variety of possibilities exist for how to make the most use of realm X. Let me present a few proposals of my own to illuminate the way to welcoming turning point P. First of all, let me explain realm X's potential contribution to a society where varieties of agriculture coexist.

Today's agriculture is truly versatile. These considerations point to some of the variables: what crops are grown and what is the size of operation? Does the farmer grow produce for home consumption or for profit? How dependent is the farmer on government subsidies? How much does the farmer rely on chemical fertilizers and agricultural chemicals? Is the operation export-oriented? Are people with physical disabilities active in farming? Does the farmer farm to find a purpose in life or to stay healthy? Does the farm promote interpersonal and social connections? Does it add greenery to urban areas? Are animal husbandry and enhancing the welfare of animals the farm's purpose?

Due to its diversity, agriculture in Japan is sometimes a source of dispute. For instance, one camp argues that theirs is the ideal form of Japanese agriculture while another does not recognize that form of agriculture as the way Japanese farming should be. To date, however, the conviction that Japanese agriculture, no matter what it is, contributes to Japan's food security provides a solid point of compromise for the contending parties, leading them recognize each other's raison d'être.

However, once turning point P occurs, bringing with it an era of surplus farmland area, the common denominator of agriculture's contribution to food security will no longer wield enough influence to accept and bind together all forms of agriculture. When this happens, it will become easier to demarcate those values that can be explained by food security and those that cannot.

For instance, measures to promote active participation of people with physical disabilities in farming, advocate farming's offer of a purpose in life and physical health, and nurture social ties can become methods to utilize realm X for social welfare. As such, all of today's current efforts can be regarded as preparatory exercises for the future. How one envisions the agriculture of the future and with that the implementation of current policies will differ greatly whether one considers social welfare-oriented agriculture as being necessary for food security, on the one hand, or advocates social welfare-oriented agriculture as a use of realm X, with more leeway from the viewpoint of food security, on the other.

The emergence of realm X supports the coexistence of agriculture in which food security plays a major role and agriculture in which food security plays only a minor role. It also provides an opportunity to positively accept the richness of agriculture, which cannot be explained by food security.

Moreover, realm X will not necessarily be monopolized by agriculture. Realm X can prompt the realization of affluence through various uses of national land and make the coexistence of various uses of the national land easier to accept.

The possibility for food and power to be 100% self-sufficient by 2100

What would happen, for instance, if solar power facilities were installed throughout all of realm X? According to a survey conducted by the Ministry of Environment in 2010, if a solar power facility is installed on a flat setting, it can generate 61.59 kW annually per square meter.

Realm X will continue to expand after the fifth turning point in 2052 to the level of 1,681,769 ha by 2099. If solar power facilities are installed to keep pace with this expansion, they alone will be able to generate power equal to that produced currently in Japan as a whole (1,024.7 billion kW), even with today's technology. If solar energy technology or energy-saving technology advances and if the demand for power decreases due to population decreases, 100% self-sufficiency for electric power may be achievable even earlier than 2099.

It will not be necessary to destroy mountain forests to install solar panels or to install them on steep slopes. Currently, the rapid of solar panels is disorderly, causing conflict and competition between some users of farmland and national land. The fact that Japan has not yet reached turning point P is a fundamental reason for these flashes of conflict and competition.

If at this point Japan could reach national consensus that solar power generation is to be conducted in the future realm X, it should be possible to gradually modify national land use in tandem with the pace of expansion of realm X and transform Japan into a country capable of self-sufficiency in both food and electric power.

Changing role of current subsidies to protect farmland

The maintenance and management of farmland, which is liable to deteriorate if left unattended, are not only significant for their roles in enhancing food security. These activities also have a role to play in expanding the space of realm X in the future (see Fig. 7.6). When a promising use of realm X for the public interest is found, it might become possible to mitigate the sense of financial burden among taxpayers accompanying subsidies aimed at preventing farmland from becoming barren or restoring farmland already in ruin. It may even become possible to increase subsidies incrementally.

For instance, if farmlands on the verge of deterioration are salvaged with a new annual subsidy of ¥1 million per hectare until a year before the fifth turning point in 2052, Japanese could help retain 6325 ha of farmlands with an additional cost of only ¥50–63 per capita per year. This means, at least mathematically, that Japan could pass on a wider realm X to future generations while at the same time protecting farmland that could supply food for an additional 150,000 people in times of emergency (see Fig. 7.7).

I need to hastily add, though, that the subsidy of ¥1 million per hectare is a mere supposition. The more farmland is maintained, the lower the subsidy can become. The closer turning point P comes to the present, and thanks to technological advancement, the shorter the duration of a subsidy will be. What's important is that Japanese farmers would receive a stable subsidy until the time limit for turning point P is reached. If turning point P coincides with the fifth turning point in 2052, the subsidy would have been provided for 30 years, which is a long enough time for a mother to bear and rear a child. The subsidy, thus provided, would not only encourage settlement in agrarian villages but also would allow people to experiment with new uses of realm X, liberated from conventional wisdom. This idea would also promote population and economic decentralization, thereby contributing to lessening the risk in times of disaster posed by excessive centralization in Tokyo.

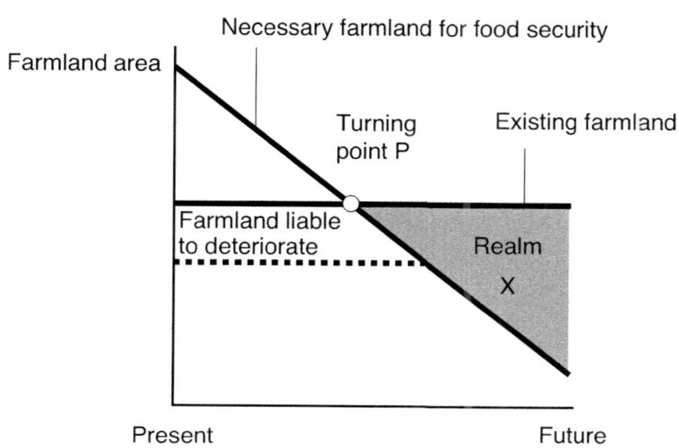

Fig. 7.6. To protect farmland liable to deteriorate is to expand future realm X.

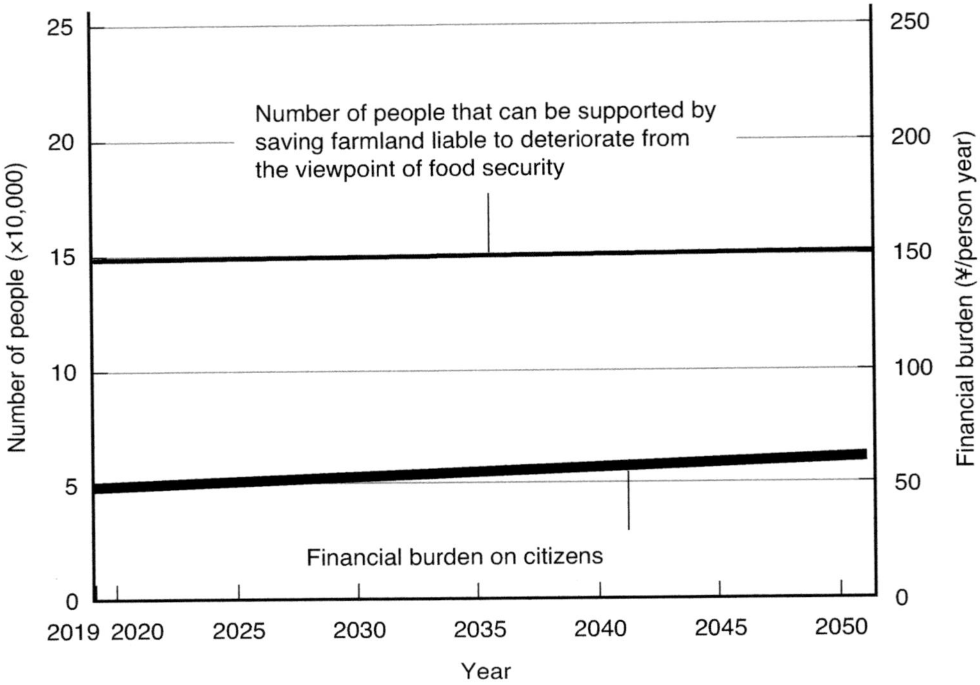

Fig. 7.7. The number of people additionally supported when farmland liable to deteriorate is saved by subsidy, and the additional financial burden on citizens. *Source*: farmland liable to deteriorate is calculated as 96% of Item (ii) Idle Cropland, as stipulated by Chapter 4 of the Crop Act, in 2019. Other figures are based on figure 3.6 in Ogawa (2022). *Note:* the subsidy amount is ¥1 million/ha.

Disbursing the kind of subsidy I am suggesting may appear reckless to taxpayers, but doing so would be much more strategic than promoting measures that cannot be fundamental remedies for long-term surpluses of farmland area. By this I mean measures to protect farmland that are based on people's obscure image of how important farmland is or attempts to construct a biotope. In short, in attending to farmland on the verge of deterioration or farmland already in ruin, it is not enough to consider what to do in the short term of this year or the next. Planning backward from the prospect of realm X is vital.

Enacting a new basic act and constitution

As the examples I've given show, there are many ways to utilize realm X. But it is my conviction that agriculture is still the best way to use it, because national land and farmland areas in Japan are extremely small relative to the population, compared to other countries, particularly western more economically developed nations. This fact makes Japan's land all the more precious. If realm X ceases to be farmland beyond the fifth turning point in 2052, farmland area in Japan in the year 2100 will be 2.51 million ha. This is a reduction of as much as 46% from farmland area in 1999 when the current Basic Act on Food, Agriculture and Rural Areas was enacted.

This 23-year-old act (at the time of writing) was enacted 38 years after its predecessor, the Agricultural Basic Act, was promulgated in 1961. Taking into consideration the obscure positioning of the importance of rice paddies and rice production as a rationale for the policy in the 1999 Basic Act and the imminent emergence of realm X, it seems high time to start studying a new Basic Act. This time, it is hoped, the new Basic Act would clarify both the importance of rice paddies and rice production as well as the future prospects for farmland and realm X.

More grandly, it also might be high time to amend the Constitution of Japan. For instance, in 2017 Switzerland amended its constitution to refer explicitly to food security. The amendment, which stipulated protection of farmland, obtained 78.7% of the vote in a national referendum.

When it comes to Switzerland, I'm sure many readers would envision the peaceful scenery of a wide expanse of grassland against the Alps mountain range. Swiss grassland is instrumental in the country's food security because it can be converted to farmland in times of emergency. In normal times, that area is used as low-cost grassland to raise farm animals. The government disburses a subsidy directly to owners of grasslands in the name of maintenance of scenery and the environment. This subsidy amounts to as much as 92.5% of the agricultural income of the owners. Swiss rural scenery has been artificially created, reflecting citizens' will and determination.

Differences in geographical conditions, histories of nation building, democratic institutions, frequency of amendments to laws, including the constitution, as well as histories of agricultural policy between Japan and Switzerland defy easy comparison between the two countries. In the case of Japan, it should be remembered that realm X is surplus farmland area from the viewpoint of food security. And yet, the Swiss experience can serve as a useful reference for Japan in strategically planning the best use of its national land and farmland area.

At present, the ruling Liberal Democratic Party advocates four major amendments to the current constitution. These have to do with the positioning of the Self-Defense Forces, a state of emergency clause, the electoral system, and education. Today, as Japan stands at the crossroads of farmland's future, it is hoped that discussions on constitutional amendments among the sovereign people of Japan would include deeper thought on food security, how best to use the national land, and what form agriculture should take in the era of surplus farmland area.

A shift in food security policy

As the era of rice shortage transformed to an era of rice surplus, rice became increasingly liberated from state control. Something similar could happen with food security. If the situation with surplus farmland area relaxes, it could trigger a transition from the era of total dependence on government subsidies for food security to a new era in which voluntary involvement in food security by citizens is nurtured while the state continues to guarantee the minimum level of food security.

Even if a consumer always buys rice directly from a farmer they've been acquainted with for years, there is no guarantee that the farmer will sell them rice in times of emergency such as a food crisis or a war. Because of this uncertainty, I propose establishing a mechanism through which it is guaranteed that rice will be delivered in times of emergency if a consumer pays an insurance premium on top of the ordinary rice price. The mechanism should also guarantee the consumer use of a portion of the rice farmer's residential land or farmland to grow their own rice. This way, consumers can secure their own food security by entering a contract with the region or rice farmer of their choice. This mechanism has the problem, however, of favoring the rich when farmland is in short supply. This mechanism would be possible only when there is surplus farmland area throughout the entire nation.

The evacuation insurance offered by Chizu-chō, Tottori prefecture, may serve as a useful reference when thinking about food security mechanisms that are supported and mediated by local municipalities. Chizu-chō's evacuation insurance provides disaster victims not only with spots at evacuation sites but also with deliveries of agricultural produce as a privilege of the insured. As part of the deal, insured persons are entitled to use municipal facilities at a reduced rate, a perk that Chizu-chō expects will promote urban–rural exchange.

When someone proactively involves themselves in food security, they engender new social ties. Particularly for dwellers of cities, where farmland is scarce, involvement in food security is an opportunity to build new relations with rural areas. The resultant social ties will greatly contribute not only to more flexible responses to unexpected as well as predicted disasters, such as an earthquake directly beneath Japan's capital of Tokyo or a Nankai megathrust earthquake, but also to resilient recovery from these disasters.

At this point, I must say that I find it particularly commendable that farmlands, and particularly rice paddies, which can grow the rice that is part of every corner of life in Japan, are, to some extent, still retained all over the country. This, indeed, is affluence.

Even when there is a surplus of farmland area, as long as people continue to be obsessed with the sense of having enough as the norm, Japan is liable to experience a poverty of invention. That is, people would continue to argue that it is enough to retain for future generations only those rice paddies that can produce tasty rice or those farmlands where crops can be grown efficiently. But it is precisely because there is a surplus of farmland area that it is important to begin to find as the norm a sense of affluence derived from overabundance that includes rice paddies that cannot produce tasty rice as well as farmlands considered to be inefficient.

Full-fledged export of agricultural produce

The words surplus farmland may bring to mind such relatively unfavorable settings as farmland situated on a mountain ridge or farmland that is difficult to handle. Depending on how one looks at it, however, it is conceivable that farmland with favorable conditions could be designated as realm X.

Let me explain. While nowadays exports of agricultural produce from Japan are increasing, it should be noted that this is occurring only in the midst of a dearth of farmland area. But if Japan were to designate efficient farmland as part of realm X and use that farmland to produce agricultural produce for export, then Japan could promote development of its export industry through farmlands on plains while securing food security by designating farmland on mountain ridges for use in times of emergency. Whether Japan can grow as a full-fledged agricultural export economy or not, over and above the level of merely promoting exports in order to secure farmland for emergency use, depends on whether the production of produce for export can be envisioned as a usage for realm X.

Exporters of agricultural produce often restrict exports, citing one reason or another. Japan can certainly relate to the pain of the victims of these export restrictions. I propose that Japan should envision development of an export industry more resilient to food crises and other contingencies by making use of realm X. Japan should do this not only for its own benefit but also for the people of importing countries.

Possibility and limit of the 'support by eating' campaign

In Japan, the catchphrase 'support by eating' is often heard. It refers to people eating their own country's produce to support farmers or eating produce from disaster-affected areas to support reconstruction efforts. But don't be mistaken: eating does not always lead to the use of farmland. Suppose one pays twice as much to buy rice in hopes of supporting rice production. A rice farmer's income will be doubled either by consumers buying twice as much rice or by consumers paying twice as much as usual for the same amount of rice.

In the context of rice paddies, however, no matter how expensive rice becomes, the amount of rice paddy area needed to grow rice does not change if the amount of rice consumed does not increase. It may be nice to think one is supporting rice production by eating rice, but it should be kept in mind that to support rice farmers is not the same as to support rice paddies.

When supporting rice production by eating more rice, it is easier to pay more for the same amount of rice than to buy more rice, given that there is a limit to the amount of rice one can eat. While paying a higher price for the same amount of produce may contribute to promotion of agriculture as an industry, it does not necessarily lead to the use of more farmland. This is even more true particularly when the population dwindles. While per-capita caloric intake has been shrinking in recent years, the national caloric intake will continue to decline due to the decrease in population even when the per-capita intake is maintained (see Fig. 7.8). The fact becomes clearer that paying a higher price for the same agricultural produce may support agriculture as an industry but cannot support retention of farmland.

If the aim is to use more farmland, then more attention should be paid to livestock. Livestock eat a higher proportion of imported agricultural produce than people do. Japan imports large amounts of overseas livestock products. Moreover, domestically produced livestock products also rely on a high percentage of imported animal feed (see Fig. 7.9). To promote more use of farmland, it will be easier to have livestock animals eat more domestically produced animal feed than for the Japanese people to eat more domestically produced agricultural produce.

How to sustain and terminate the subsidy for rice for animal feed

The observation just expressed notwithstanding, I am not arguing that the current government subsidy for rice for animal feed be sustained. The government now disburses a maximum ¥1.05 million per hectare as a basic subsidy for rice for animal feed. As explained in Chapter 6, this outrageous amount is attributable to the problematic design of the subsidy. Here I wish to propose the following two remedies.

First, the subsidy for rice for animal feed should be split into two: (1) a high but temporary subsidy provided on the condition that the recipient is committed to producing dry upland crops in the future and (2) a low but permanent subsidy.

In rice paddies that grow highly competitive brands of rice for meals, rice farmers are liable to refrain from growing rice for animal feed because there is not much benefit in producing dry upland crops or rice for animal feed with the meager subsidy. But the temporary subsidy that I propose would allow farmers to grow rice for animal feed as a stop-gap measure in rice paddies that cannot be suddenly switched to dry upland crops. With the subsidy, farmers could secure both the preparation time and the funds to shift to growing dry upland crops.

Those rice paddies receiving the low but permanent subsidy for rice for animal feed would be mostly those which cannot produce any other promising produce than rice, or those a farmer has chosen to leave temporarily uncultivated.

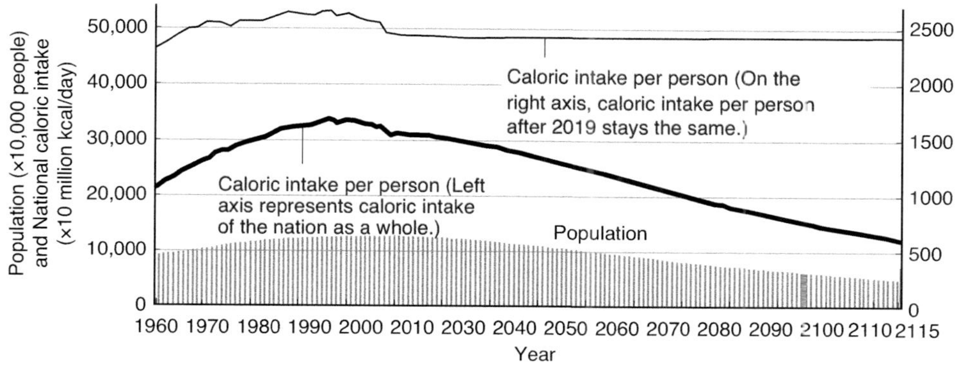

Fig. 7.8. Changes in caloric intake and population. *Source*: based on the Ministry of Agriculture, Forestry and Fisheries' *Shokuryō Jukyūhyō* [*Food Supply and Demand Balance Sheet*], the Ministry of Internal Affairs and Communications' *Jinko Suikei* [*Population Estimates*], and the National Institute of Population and Social Security Research's *Nippon no Shorai Suikei Jinko (2017-nen Suikei, Shussei Chui-Shibo Chui)* [*Population Projection of Japan (2017 projection of medium-fertility and medium-mortality)*].

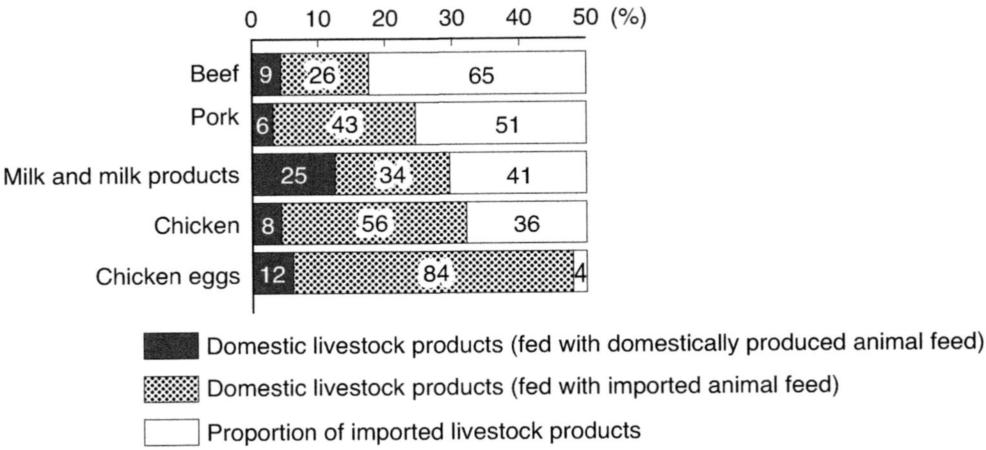

Fig. 7.9. Proportions of domestic livestock products and imported livestock products by type, fiscal year 2019. *Source*: based on the Ministry of Agriculture, Forestry and Fisheries' *Shokuryō Jukyūhyō* [*Food Supply and Demand Balance Sheet*].

Farmers with rice paddies that can only produce rice and which without the subsidy for rice for animal feed would deteriorate would be particularly inclined to choose the low but permanent subsidy.

In the end, it is rice farmers who grow rice to be eaten by family members and relatives, aside from those rice farmers who can produce highly competitive brands of rice, who prefer to grow rice for meals over rice for animal feed. When rice for meals is produced by these farmers who can endure low prices, consumers will benefit by lower rice prices and fiercer competition among rice-producing regions. Those farmers who produce highly competitive brands of rice will waste no time to explore new opportunities that utilize their advantages, such as exports, instead of being lured by a subsidy into growing rice for animal feed.

The second remedy I propose is to attach conditions to the production of rice for animal feed. For instance, under current policy, a subsidy is provided equally for rice for animal feed grown with chemical fertilizer alone and that grown with compost, a practice that contributes to resource recycling. If the subsidy for rice for animal feed is offered exclusively to rice farmers using compost, it becomes clear that production of rice for animal feed contributes to easing the strain on the environment and makes a significant social contribution. The result would be better understanding of the subsidy for rice for animal feed among the Japanese people. This remedy means that rice farmers who cannot procure compost cannot grow rice for animal feed. But it would allow the subsidy to be concentrated on those rice paddies that have to rely on growing rice for animal feed even if the owners have to go through the trouble of somehow procuring compost.

Rural areas where rice whole crop silage becomes rampant

What I personally consider more promising than rice for animal feed is the whole crop silage (WCS) method for rice, or RWCS. This method can use a broader range of rice paddy area than rice for animal feed while the necessary labor time remains the same. Because income from its sale per hectare is higher than rice for animal feed, RWCS also offers more leeway to reduce subsidies and improve technology.

As explained in Chapter 5, while RWCS is cumbersome to handle, it is a relatively easy operation to sustain once started. Unlike rice for animal feed, which retains the same quality no matter who produces it, the quality of RWCS reflects a farmer's skill. Therefore, farmers with rich experience in RWCS can market their products even to other regions.

Producing RWCS does not require a rice dryer. As explained in more detail later, however, because RWCS produces fewer kernels, it is more difficult to convert it to rice for meals in times of emergency. On the other hand, from the viewpoint of the usage of a realm X liberated from the food security requirement, production of RWCS is a highly promising option that can maximize realization of the multifunctional roles of rice paddies and rice production as well as the use of large rice paddy areas to produce animal feed with smaller subsidies.

The special breed of rice exclusively for RWCS consists of stalk and leaves that are easily digested by cattle and fewer kernels, which are indigestible. The breed Tsukisuzuka, for instance, grows to 121 cm tall but it produces only 0.5–1 ton of kernels per hectare. This contrasts sharply with the Koshihikari breed, which is about 80–90 cm tall but produces as much as 5 tons of kernels per hectare. Production of RWCS is truly production of rice for animal feed.

In order to transport RWCS cheaply and upgrade the animal feed it is processed into, more efficient methods of RWCS production are being studied. These include ways to store massive amounts of unrolled RWCS and wider use of a machine that enables delicate cutting of the rice plant.

Roughly speaking, animal feed for cattle is divided into two types: coarse feed that is high in fiber including RWCS and concentrated feed that is high in nutritional content such as rice for animal feed. While it is easy to obtain coarse feed in Hokkaidō through cultivating grassland and pasturing, other prefectures rely heavily on imported cured hay. The fact that rice production is becoming increasingly popular in Hokkaidō while rice production in other prefectures is lessening means that producing RWCS in rice paddies in prefectures other than Hokkaidō is a highly promising use of the national land for the future.

At present, the production of RWCS is an attempt to make farmland cater to one's stomach, so to speak, meaning the stomachs of livestock. The more RWCS is produced, the more livestock will eat RWCS and, as the Japanese become fond of the taste of livestock fed with RWCS, its production will become a promising use of realm X.

Originally, cattle ate coarse feed. As cattle became domesticated, however, they were fed increasingly with concentrated feed. Particularly in Japan, where animal husbandry was developed on the premise of heavy reliance on imported animal feed, all the prefectures, including Hokkaidō, experienced development of animal husbandry reliant on imported concentrated feed. If the proportion of course feed, which is the original, natural food for cattle, increases, there will be more room to increase the domestic production of pasture grass and RWCS.

A domesticated animal's growth and body structure differ in accordance with the type of feed the animal receives and the timing of the feeding. So does composition of the animal product. Studies have been conducted on how the quality of the animal product changes when the animal is fed with plenty of grass and RWCS. Results vary, ranging from 'no tangible change' to 'tastier' and 'changes in scent and appearance.' It should be noted that the relation between the tastiness of beef and the components of the meat has not yet been fully clarified scientifically. It should also be noted that whether meat tastes good or not is different with each person. Even for the same person, taste changes with age. As one grows older, for instance, red meat may taste better than marbled beef. A person may also become increasingly health conscious, which may affect their choice of meat. It is known that milk from cows fed with an abundance of summer green grass tends to smell like grass. Over time, an increasing number of Japanese may grow to like it as a more natural taste of milk.

In the future, the trinity of people's stomachs, the stomachs of livestock, and how farmland is used may accomplish a dynamic transformation beyond the shortsighted choice of whether to make stomachs cater to farmland or vice versa.

Tsukikotoka, a rice breed specially developed for RWCS, grows taller than 150 cm. (Photo supplied by the National Agriculture and Food Research Organization. Used with permission)

Increasing citizens' burden and benefit

In the past and now, ordinary consumers have supported Japan's agriculture by buying its products and paying taxes. And the support of domestic agriculture has been directly linked to the use of domestic farmland. In this era of surplus farmland area, however, prosperity for agriculture does not necessarily lead to solution of the problem of surplus farmland. This is the same structure inherited from the age of surplus rice paddies when development of the rice industry did not contribute to solution of the rice paddy surplus problem. The arrival of turning point P will make people realize that to support agriculture alone is no longer sufficient for obtaining affluence generated by farmland.

It would be best if the use of realm X for agriculture could be done in a way that is economically viable without relying on government subsidies. This way, multifunctional roles of agriculture and rural areas can be fully realized through maintenance of agriculture in realm X without relying on government subsidy. In actuality, however, it will be difficult for agriculture to use up the entire realm X without subsidy. Besides, as the population of Japan continues to decrease, so will realm X per-capita increase. In 2054, 2 years after the fifth turning point, realm X per capita will be only 0.09 ares, but it will increase to one are per capita in 2073, 2 ares in 2089, 3 ares in 2102, and to over 4 ares per capita in 2114.

One may wonder how much financial burden the Japanese people are willing to bear for realm X. As indicated in the section of Chapter 6 entitled 'Problems with Current Policies,' the monetary value of the multifunctional roles of agriculture and rural areas was ¥1,928,043 per hectare for rice paddies and ¥691,826 per hectare for cultivated dry fields, exclusive of such factors as food security, agrarian culture, and biodiversity.

Let us assume that when realm X is no longer farmland, the benefits of a subsidy will no longer be available. When a subsidy of ¥650,000 per hectare is provided to realm X as farmland, however, the benefit it engenders for Japan supersedes the financial burden for its people if the subsidy is used for cultivated dry fields. The benefit will supersede the burden even more if the subsidy is used for rice paddies.

When ¥650,000 per hectare of subsidy is disbursed annually for farmland in realm X after the fifth turning point of 2052, the total additional financial burden on the Japanese people will be ¥472.3 billion in 2070 and ¥1111.4 billion in 2100 at today's value. Those amounts constitute 0.46% and 1.36%, respectively, of the 2019 tax revenue of ¥103,386.6 billion, which is the sum of national and local taxes. This will be an annual financial strain of ¥5674 per capita in 2070 and ¥18,075 in 2100 for the Japanese people. If this strain allows us to use realm X for agriculture, I personally think it is a good deal. I wonder how my readers will judge.

What I would like to see is for these concrete figures to trigger discussions on realm X. For instance, as pointed out in Chapter 6, the multifunctional roles of agriculture and rural areas have provided an obscurely perceived policy rationale. If this policy rationale continues to be passed on as is in days to come, it follows that, as the population of Japan continues to decrease, the per-capita benefit from agriculture and rural areas will continue to increase. What is imperative now is more rigorous discussion of policy rationale from the perspective of whether each Japanese can really benefit more from realm X and how new benefits can be found.

Also, the more discussion there is, the more the goal of Japanese agriculture will be clearer. For instance, if a social consensus is formed that paying ¥3000 per capita in support of agriculture is permissible in 2070, a goal can be set up for technological advancement and cost reduction. In agricultural policy, easy-to-understand goals tend to be set, including such quantitative goals as doubling income, cutting production costs by half, and even 'increase no matter what, reduce no matter what.' But the rationales behind these goals are not at all concrete. By focusing attention on turning point P and a gradually incremental realm X, I believe it will become possible to set strategic goals and adjust and verify goals in response to the actual situations regarding farmland and population.

Economical way to face realm X

Some readers may find my trial calculation that each person is expected to bear an annual burden of ¥5674 in 2070 and ¥18,075 in 2100 to use realm X for agriculture to be too heavy a responsibility, even taking into consideration the possibility of a lesser burden thanks to technological advancement or the rise of new businesses in realm X that would make it unnecessary to rely on subsidies. Let me, then, propose a more economical alternative. That is for Japanese to give up altogether on making a plan for realm X. Once realm X is left to the weeds, trees, birds, beasts, and other animals, the farmland will go barren in no time.

Realm X could be entrusted to other entities besides wild animals and plants. Entrusting realm X to foreign capital could also be promising. When a law was enacted in 2021 to regulate the use of important national land for the security of Japan, including national defense facilities and nuclear power plants, no consideration was given to food security. Also, there is no fundamental rule in Japan to regulate the buying and selling of real estate by foreign capitalists. As a matter of fact, significant and increasing numbers of foreign capitalists have been purchasing sites for solar power and wind power generation; they've also bought mountain forests, which contain sources of water. Foreign capitalists have already targeted farmland, too. In 2020, a corporation in which 49% of voting control was owned by a Chinese–Hong Kong corporation obtained 10.1 ha of farmland in Saijō-shi, Ehime prefecture. This is only part of the total of 65.9 ha of farmland purchased by foreign capitalists from 2017 to 2020, and farmland area bought by foreign capitalists continues to grow steadily. At present, the Cropland Act prohibits a corporation whose majority of voting rights are not held by those engaged in agriculture to own farmland. Recently, however, there have been movements to relax the regulation to allow corporations led by non-agricultural capitalists, including overseas corporations, to own farmland.

The prospects for realm X are an excellent theme in the exploration of the future of the Japanese people. If the future of realm X is entrusted to foreign capitalists, Japanese will no longer have to think about what will become of realm X. The argument that surplus farmland area should be returned to nature sounds comfortable to Japanese given how they have cohabitated with nature. Besides, it is also a highly economic option. Giving up on farmland, however, will not turn the clock back to simpler times. And, in fact, it is quite possible that doing so would mean that the use of Japan's national land is determined by foreign hands. Once that happens, it will be impossible to turn the clock back from that point to the present. To plan what to do with surplus farmland is to plan the future.

In other words, bearing a financial burden to provide a subsidy for realm X represents a responsibility for the Japanese people to keep realm X in their own hands. If realm X can be considered national land to be protected, it might be permissible to a certain extent for Japan to tolerate a subsidy for realm X as a de facto defense budget, seeing as the per-capita annual defense budget now is only ¥49,000, or only 20% of that of the United States and about 40% of that of South Korea and Australia.

Whether the Japanese people feel the financial burden imposed on them incurred by the protection of farmland with government subsidies before the arrival of turning point P and after that for realm X to be heavy or economical depends greatly on how the Japanese face realm X.

Time to thoroughly review perceptions of agriculture, farmland, and national land

Can the Japanese people obtain more affluence from farmland than before? Particularly, how much affluence beyond just food security the Japanese people can envision in realm X directly links to the future of agriculture, rural areas, and national land use in Japan. Japanese will be tested on not how they evade the greatest crisis called realm X but how they accommodate this crisis and overcome it.

One cannot discuss the rice problem, including the rice paddy problem, by paying attention only to the amount and price of rice. Similarly, the issue of the future of agriculture, including farmland and national land use, cannot be explored by only focusing on the amount and price of agricultural produce. The era when the future of rice paddy area can be discussed by considering only the future potential of the rice industry is long gone. The arrival of the next turning point is imminent. When it comes, the era in which one can talk about the future of farmland on the basis of the future potential of agriculture will end.

What has become important in this situation is, in my opinion, the subjective view of each individual Japanese. I cannot help but think that now is a time when every Japanese is being asked to look more deeply at the significance and values they find in rice production and agriculture as part of the overall picture of agriculture, farmland, and national land use in Japan. At this point in time, the subjective view of every Japanese is being tested.

Note

[1] Although some argue that farmland would not be deficient if it were mainly planted with potatoes, not enough attention is paid to wet paddy area when calculating the food self-sufficiency potential index. Farmland in Japan is, indeed, deemed deficient (see Ogawa, 2022).

References

Nihon Keizai Shimbun (2015) *Eiyō Katayori Kadai ni* [*Nutritional Imbalance Becomes an Issue*]. March 18. Nihon Keizai Shimbun, Tokyo.

Ogawa, M. (2022) *Gendai Nihon Nōgyō Ronkō* [Ideas for Contemporary Japanese Agriculture]. Shunpūsha, Tokyo.

Shokuryō Anzen Hoshō ni Kansuru Kenkyūkai (Study Group on Food Security) (2010) *Wagakuni no Shokuryō Anzen Hoshō eno Atarashii Shiza* [*A New Viewpoint on Japan's Food Security*]. Translated from the Japanese original by Makito Noda.

Epilogue

My interest in the rice problem in Japan dates back to more than 25 years ago. In those days, I walked a 6 km round trip to elementary school. On the way to school, I saw nothing but rice paddies, mountains, and farmhouses. As I walked, I liked to contemplate this and that about rice.

What caught my attention were the changes and differences evident as rice grew and in how rice paddies were used. Looking closely at the rice paddies, I noticed the many changes that occurred every year, as well as season-to-season changes. For instance, from planting to harvest I noticed changes in the distance between rice stubbles, the machinery used, and the farmers driving the machines. Even when the same breed of rice is planted, the color of the leaves and the growth of weeds differ significantly depending on the farmer and the paddy. And the condition of rice paddies can differ greatly. Some rice is grown in apparently narrow rice paddies with unfavorable conditions, while other rice grows in spacious, more favorable-looking paddies. These latter paddies, however, can fall into disrepair abruptly from one year to the next. When that happens, they may be planted with eggplant or cosmos as if to provide an alibi, or converted to residential sites.

My daily commuting route was, thus, full of changes to be noticed if I paid attention. It was an excellent classroom for me more than anything else. And as I walked on the same road for 6 years, I discovered that when adults talked about rice, at times they might be talking about rice per se and at other times they might actually be talking about something related to rice paddies or farmland. I supposed that matters were much more serious when the talk was about paddies or farmland. That insight I had long ago as a boy stayed with me and became something I cherished. Having held onto that insight, over the years I began to think that if only the blessings from rice paddies could be multiplied, rice paddies could contribute to solving some of the gravest social issues.

My grandfather was a master hand at straw work. When I was in eighth grade, I asked him to weave some *shimenawa*, the hemp rope that is a familiar adornment of the Shinto religion used during the New Year holidays. I sold the *shimenawa* by myself from a stall that I had rented near the train station. That is a fun memory. It was funny that, at the time, Grandpa insisted that we should draw up a contract between us. Ordinarily, a grandfather would help his grandson free of charge, but mine was no ordinary grandfather. In retrospect, maybe he wanted to teach me how to assess and calculate materials costs, labor, and commission fees. Probably he also wanted to take time to advise me that I should negotiate with the owner of the stall with an offer to pay 10% of the sale. Grandpa taught me many things. How to weave *shimenawa*, of course, but also where to find *urajiro* (*Gleichenia japonica*) and *yuzuriha* (false daphne or *Daphniphyllum macropodum*), two of the materials for *shimenawa*. I remember him guiding me across pathless mountain slopes and snow-covered thickets as we looked for these plants. During these explorations, Grandpa taught me that the most important thing was to be familiar with the land.

I earned about ¥80,000 in the last 3 days of the year, and I did the same the following year. Although the sum was nothing more than a little extra spending money, selling *shimenawa* became my first important business career in the sense that I had earned money while attempting to find a solution for a social issue in my own way. Having gotten a taste of success, I went on in high school to cultivate *ebisugusa* (*Senna obtusifolia*) on a rice paddy slope to produce tea leaves to sell. I also studied methods to grow lotus root. While I cannot say that I devoted myself completely to preparing for the college entrance exam, I decided to study in the university agricultural department to deepen my knowledge about rice production and rice paddy usage.

Subsequently, in reading scholarly works and academic treatises on Japan's agriculture and rice, I noticed that these publications are full of such expressions as "shift," "change," and "conversion."

Agriculture and rice production have been with Japan since ancient times. Therefore, it is unconceivable that Japan will face such a dramatic change as the extinction of agriculture or rice production. What scholars refer to as a shift (or change or conversion) is, so to speak, a newly added geological layer. Past history forms a multi-layered thickness, to which new layers are added. The turning points of the rice problem do not necessarily resolve all past rice problems. Covered up by newly added layers, all past problems have been passed on until today. And, just as geological layers differ from region to region, the problems of agriculture and rice differ from region to region in accordance with respective histories, geographical, and social conditions, and individuals and communities.

Scholars must have chosen to use the term *shift* frequently because they have devoted their time and energy to the discovery and analysis of these so-called geological layers. As if studying the regional distribution of the layers, those scholars must have studied the history of changes in rice production based on, for instance, the records of rice farmers from the Edo period (1603–1868) excavated from antique documents and the evolution and diffusion of rice production technologies. Scholars have also studied changes in legal systems, which, while at a glance appearing to affect the entire nation like volcanic ash covering all land after a massive eruption, have in actuality affected different regions and individuals in significantly different ways. In certain regions, the history of rice production has been dug out like an exploratory boring. Like geological layers, the rice problem exhibits both spatial and temporal expansion. As time goes on, the areas under study will continue to widen to reveal numerous new turning points.

Yet scholarly interest in agriculture is not necessarily uniform. There has been a broad accumulation of studies on rice, in particular, and even the smallest differences have been looked into. In addition, rice studies are full of jargon that defies easy mastery; so much so that I did not know where to start when I was around 20 years old. In those days, I strongly wished for an introductory book on the rice problem that even an upstart like myself could read easily.

It is with that miserable memory of struggling to make sense of what I was reading that I authored this book. In a reversal of the usual academic approach of exploring and analyzing turning points, I decided to see how many references to turning points I could eliminate to explain the rice problem. I rigorously selected only those essential turning points that should be considered basic knowledge – or educational common sense, if you will – necessary to be passed on to future generations. I attempted to explain the rice problem in plain language, avoiding technical terms as much as possible.

Issues such as dealing with rice surpluses through the policy to reduce rice paddy area, the rice price, scales of business operation, the lack of successors, ageing farmers, and abandoned farmland are often highlighted regarding the rice problem. These individual issues are important, yes. What is most important, though, is to learn about the turning points introduced in this book more than anything else.

Growth in rice production as well as the number of rice farmers in Japan has not been straightforward. The years have seen rises and declines in both. Also, rice problems have appeared in great variety in different times and different regions. Analyzing the rice problem from the broad perspective of rice shortages through a rigorous selection of past pivotal points, I found that 1967, a half century ago, was the most significant turning point in the history of rice production in Japan since the Jōmon period (*c.*14,000 to 1,000 BC). On the basis of this discovery, I succeeded in understanding subsequent turning points, resultant rice problems, and the gravest crisis about to confront Japan.

This book was published originally in Japanese on June 25, 2022. That was the year of the great volcanic eruption in Tonga in January and Russia's military invasion of Ukraine in February that triggered fears of a global food crisis. Worry over a food crisis can spark renewed appreciation of a country's agriculture. For Japan, the importance of the luxury of a stable supply of rice, regardless of circumstances worldwide or a global hike in cereal prices, will be appreciated once again. Anxiety about a global food crisis is a tailwind for Japan's agriculture and, particularly, rice production. It is all the more imperative now that Japan pay closer attention to the gravest crisis currently creeping along below people's awareness.

The theory on turning point P and realm X as the gravest of crises for Japan, introduced in Chapter 7, is also an elegy or requiem on the passing of rice paddies and farmland, which had for so long been pillars of support for the meals so familiar to Japanese, one after another as unwanted facilities. The theory is also a celebration of surplus rice paddy and farmland areas as symbols of the Japanese people's affluence. Of these two faces to the theory, I personally hope that the face to be accepted by Japanese is the one that celebrates surplus rice paddies and farmlands.

To celebrate surplus rice paddy and farmland areas is to positively acknowledge the reality of their existence. This means accepting all of their properties, including the perceptions that they are a waste of space and have a fragile existence reliant on government subsidy but also the possibility that they can be used in ways that differ from traditional agriculture. What is at stake at this stage is a courageous and resolute decision on how to handle realm X based on the vision Japanese hold for the form agriculture, farmland, and national land use should take, detached from the conventional vision of agriculture as a means to obtain food and the protection of farmland as a means to ensure food security.

Each individual Japanese, I believe, has to take up this challenge. It has often been argued in recent years that how far rice paddy and farmland area should be protected must be based on a strategy determined by the farmers and citizens of each region. While this argument may sound as if it respects the self-initiative of each community, it is tantamount to those outside a community merely imposing the burden of solving the problem on the current generation within a community actively engaged in agriculture. Expecting small communities of farmers and citizens to come up with a future vision for the continuously growing rice paddy surplus and the imminent problem of surplus farmland area has already nearly reached the limit. Coexisting in this narrow national land as the Japanese do, it is wrong to weigh down only one portion of the population with the responsibility for finding ways to handle the rice paddy surplus and realm X. It is imperative for each and every Japanese to give serious thought to how to pass on our wisdom on what to do with the issue of the rice paddy surplus and realm X to future generations. I would be delighted if this book becomes a spearhead to this discussion.

Bibliography

Aikawa, T. (1994) *Kome Gyōkai wa Yami no Naka* [*Rice Industry is in Darkness*]. Daiamondo-sha, Tokyo.

Akaneya, T. (2003) Hi-Bōekiteki Kanshin Jikō no Seijigaku [Politics of non-trade concerns]. In: Kodera Akira (ed.) *Tenkanki no WTO* [*WTO in Transition*]. Tōyō Keizai Shimpōsha, Tokyo, pp. 47–85.

Andō, H. (2014) Suiden Chūshin Shikan Hihan [Critique of rice paddy–centered historical view]. *Kokuritu Rekishi Minzoku Hakubutsukan Kenkyū Hōkoku*, 185, 405–448.

Arahata, K. (2014) *Gentan 40-nen to Nihon no Suiden Nōgyō* [*40 Years of the Acreage Reduction Program and Rice Paddy Agriculture in Japan*]. Nōrin Tōkei Shuppan, Tokyo.

Arahata, K. (2018) Kome Seisan Chōsei no Kuni karano Haibun Haishi wo Meguru Nihonshi-teki, Sekaishi-teki Taikyokukan [A wider perspective from the Japanese and world history viewpoints on abolition of rice rationing policy in the country of rice production adjustment]. *Nōgyō*, 1635, 4–5.

Conlon, M. (2008) *Japan. Agricultural Situation: The History of U.S. Exports of Feed Grains to Japan*. GAIN Report (JA8521). USDA Foreign Agricultural Service.

Dower, J.W. (1999) *Embracing Defeat: Japan in the Wake of World War II*. W. W. Norton & Company, New York.

Endō, T. (1961) "*Tōhoku Kome no Kai" no Kiroku* [Record of Tohoku rice society]. In: Nōsei Chōsa Iinkai (ed.) *Nihon no Nōgyō*, vol. 4. Agriculture in Japan, Tokyo.

Endō, T. (1964) (supervised by Sagae Kin'ichi). *Sengyōsha no Kome Tsukuri* [*Rice Production by Full-time Farmers*]. Nōsei Chōsa Iinkai, Tokyo.

Fuchino, Y. (2000) Yūkyū Nōchi no Jittai to Kaishō heno Kadai [Actual situation of idle farmland and obstacles to resolution of the situation]. *Nōgyō Jōhō* 487, 2–62.

Hata, M. (1984) Kengyō Nōka [Part-time farmers]. *Nōgyō to Keizai* 50(14), 94–102.

Hirose, M. (1981) *Hojokin to Seikentō* [*Government Subsidy and the Ruling Party*]. Asahi Shimbunsha, Tokyo.

Ikeyama, K. (1986) Jimintō Asshō de Sengo Seiji no Sōkessan ni Jishin Tsuyomeru Nakasone Shushō [Prime Minister Nakasone becomes more confident in taking a thorough inventory of the postwar politics with LDP's landslide victory]. *Nihon Nōgyō no Ugoki* 79, 4–21.

Imamura, N. (1978) *Hojokin to Nōgyō/Nōson* [*Government Subsidy and Agriculture/Agrarian Villages*]. Ienohikari Kyōkai, Tokyo.

Ino, H. (1978) *Moshihogusa* [*Eelgrass*]. Gyozankai, Tokyo.

Ishii, H. and Yamamoto, S. (1987) Kantō Chihō no Daichi Riyō ni okeru Rikuden no Igi [Significance of cultivated field with water supply in the use of plateau plain in Kanto area]. *Tsukuba Daigaku Jinbun Chirigaku Kenkyū* 11, 1–19.

Jinkō Mondai Shingikai (ed.) (1974) *Nihon Jinkō no Dōkō* [*Trends in Japan's Population*]. Okurashō Insatsukyoku, Tokyo.

Kaikeikensain. (2016) Kome no Seisan Chōsei Taisaku no Jisshi Jōkyō to ni Tsuite [On status of implementation of rice production adjustment measures]. Office of Audit of Japan, Tokyo.

Kajii, I. (1973) *Shō-Kigyōnō no Sonritsu Jōken* [*Conditions for Survival for Small-scale Farming Corporations*]. University of Tokyo Press, Tokyo.

Kajii, I. (1986) *Gendai Nōsei-ron* [*Contemporary Theory of Agricultural Policy*]. Kashiwa Shōbō, Tokyo.

Kajii, I. (1997) *Kokusaika Nōsei -ki no Nōgyō Mondai* [*Agricultural Problems in the Era of Internationalization of Agricultural Policy*]. Ienohikari Kyōkai, Tokyo.

Kanazawa, N. (1989) *Suiden Nōgyō wo Kangaeru* [*Thoughts on Rice Paddy Agriculture*]. University of Tokyo Press, Tokyo.

Karube, K. (1997) *Nichibei Kome Kōshō* [*US–Japan Rice Negotiation*]. Chūō Kōronsha, Tokyo.

Kashiwagi, M. (1994) *Gendai Chūsankan Chiiki Nōgyō-Ron* [*Contemporary Theory of Agriculture in Mountains and Hilly Areas*]. Ochanomizu Shōbō, Tokyo.

Kashiwagi, M. (2002) *Jōken Furi Chiiki Saisei no Ronri to Seisaku* [*Logic and Policy of Rehabilitation of Areas with Unfavorable Conditions*]. Nōrin Tōkei Kyōkai, Tokyo.

Kawada, S. and Yamazaki, K. (1976) Saibai Kanri niyoru Chiryoku Baiyō [Restoration of soil fertility through cultivation management]. In: Nōsei Kenkyū Center (ed.) *Nippon no Chiryoku* [*Soil Fertility of Japan*]. Ochanomizu Shōbō, Tokyo, pp. 134–158.

Keiō Gijuku Shuppansha (ed.) (1960) "Nōgyō wo Ronzu" [Speaking of Agriculture] in *Jiji Shimpō*, May 2, 1883. In: *Fukuzawa Yukichi Zenshū* [*Complete Works of Fukuzawa Yukichi*] (compiled by Keiō Gijuku), vol. 8. Iwanami Shoten, Tokyo, pp. 640–646.

Kikuchi I. (2019) *Ue to Shoku no Nihonshi* [*History of Starvation and Food in Japan*]. Yoshikawa Kōbunkan, Tokyo.

Kim Chan, D. (1999) Kome Shijō Kaihō ni Kansuru Seisaku Kettei Katei no Gyōsei Sekininron teki Bunseki [Analysis of administrative responsibility in policy decision process toward opening of rice market]. *Hōngō Hōsei Kiyō* 8, 187–233.

Kitamura, J. (1980) Suiden Riyō Saihen to Tochi Kairyō Jigyō [Rice paddy usage reorganization and land improvement measures]. *Nōgyō Doboku Gakkaishi* 48(9), 633–638.

Kitō, H. (1983) *Nihon Nisennen no Jinkō-shi* [*2,000-year History of Japan's Population*]. PHP Kenkyūsho, Tokyo.

Kobori, K. (2011) *Ise Jingū no Kororo: Shikinen Sengū no Imi* [*Sprit of the Ise Shrine: Meaning of Grand Relocation of the Ise Jingu shrine*]. Tankōsha, Tokyo.

Kōchi Shimbunsha (ed.) (1970) *Inasaku Kakumei* [*Rice Production Revolution*]. Kōchi Shimbunsha, Kōchi.

Kōdansha (ed.) (1989) *Shōwa Dai 7-kan* [*Shōwa Volume 7*]. Kōdansha, Tokyo.

Kokumin Keizai Kenkyūsho (ed.) (1959) *Nihon no Shokuryō* [*Food of the Japanese People*]. Tōyō Keizai Shimpōsha, Tokyo.

Kondō, Y. *et al.* (eds.) (1979) *Nihon Nōsei no Tenkan* [*Transformation of Japan's Agricultural Policy*]. Ochanomizu Shōbō, Tokyo.

Kondō, Y. *et al.* (eds.) (1982) *Kihonhō Nosei no Sōtenken* [*Thorough Review of Agricultural Policy Under Basic Act on Food, Agriculture and Rural Areas*]. Ochanomizu Shōbō, Tokyo.

Kōnosu, K. (1981) Nōseishin Tōshin no Nerai to Tokuchō [Aims and characteristics of the Agricultural Policy Council's final report]. *Nihon Nōgyō no Ugoki* 57, 18–38.

Kumano, T. (2021) *Burando Mai Kaihatsu Kyōsō* [*Brand Rice Development Competition*]. Chūō Kōron Shinsha, Tokyo.

Marton, R. (2014) *Rice: A Global History (Edible)*. Reaktion Books, New York.

Ministry of Foreign Affairs (n.d.) *GATT/Uruguai Raundo (Nōgyō)* [GATT Uruguay Round (Agriculture)], archival document collected from Diplomatic Archives of the Ministry of Foreign Affairs, portions open to public April 2022 of 2019-100701009 and 2020-0784–0787.

Mitsubishi Sōgō Kenkyūsho (1991) *Suiden no Motarasu Gaibu Keizai Kōka ni Kansuru Chōsa/Kenkyū Hōkokusho* [*Research Report on External Economic Effect Brought about by Rice Paddies*]. Mitsubishi Research Institute, Tokyo.

Miura, H. (2012) Uruguai Raundo Kōshō ni okeru Nihon no Taiō [Japan's responses during the Uruguay Round negotiations]. *Hōsei Ronsou* 48(2), 128–151.

Nakagawa, M. (1993) Amerika no Kome Jukyū Chōsei Seisaku [Rice supply and demand adjustment policy of the United States]. *Nōgyō Sōgō Kenkyū* 47(1), 59–138.

Nakano, K. (1983) Rinchō Tōshin to Nōgyō/Nōsei no Kadai [Final report of the Ad Hoc Commission on Administrative Reform and problems of agriculture and agrarian policies]. *Tochi to Nōgyō* 23, 17–33.

Nakayasu, S. (1996) *Seisan Chōseika no Nōgyō Kōzō* [*Structure of Agriculture Under the Production Adjustment Measures*]. Nōrin Tōkei Kyōkai, Tokyo.
Nihon Gakujutsu Kaigi (2001) *Chikyū Kankyō & Ningen Seikatsu ni Kakawaru Nōgyō oyobi Shinrin no Tamenteki na Kinō no Hyōka ni Tsuite (Tōshin)* [A Report Submitted by the Science Council of Japan on Evaluation of Multifaceted Functions of Agriculture and Forest which affect the Earth's Environment and People's Life]. The Science Council of Japan.
Nihon Keizai Shimbun (2014) *Kome Seiiki wa Mamoretanoka 5* [*Has the Sanctity of Rice Been Successfully Defended? Part 5*]. Nihon Keizai Shimbun, Tokyo.
Nihon Nōgyō Kenkyūsho (ed.) (1994) *Dokusho ga Ataeta Kome no Atarashii Michi* [*New Path for Rice Provided by Book Reading*]. Nihon Nōgyō Kenkyūsho, Tokyo.
Nippon Nōgyō Kenkyūsho (ed.) (2017) Heisei 29-nendo Nippon Nōgyō Kenkyūsho Kōenkai Kiroku [Record of lectures at Nippon Nōgyō Kenkyūsho in 29th year of Heisei]. In: *Nōgyō Kenkyū* (special edition no. 5). Nōgyō Kenkyū, Tokyo.
Nippon Nōgyō Kenkyūsho (ed.) (2018) *Kome Seisaku no Minaoshi ni Kansuru Kenkyū* [*A Study on Reevaluation of Rice Policy*]. Nippon Nōgyō Kenkyūsho, Tokyo.
Noma, H. (2001) Kandenka [Conversion to dry paddies]. In: Arizono, S., Kendō, T., Onodera, A. *et al.* (eds.) *Rekishi Chiri Chōsa Handbook* [*Handbook on History and Geography*]. Kokin Shoin, Tokyo, pp. 73–78.
Nōrin Suisanshō Nōgyō Sōgō Kenkyūsho Nōgyō/Nōson no Kōeki Kinō no Hyōka Kentō Team (1998) Daitaihō ni yoru Nōgyō/Nōson no Kōekiteki Kinō Hyōka [Evaluation of agriculture and rural areas' public interest functions by alternative method]. *Nogyo Sogo Kenkyu* 52(4), 113–118.
Nōsei Chōsa Iinkai (1975) *Nōgyō Tokei Yōgo Teigi Soran* [*Comprehensive Inventory of Definition of Agricultural Statistical Terms*]. Nōsei Chōsa Iinkai, Tokyo.
Nōsei Chōsa Iinkai (1983) *Kome Ryūtsū Kōzū no Henbō to Chiiki Ruikei* [*Transformation of Rice Distribution Structure and Regional Typology*]. Nōsei Chōsa Iinkai, Tokyo.
Nōsei Chōsa Iinkai (ed.) (2021) *Kome Sangyō ni Mirai ha Aruka* [*Is There a Future for the Rice Industry?*] Nōsei Chōsa Iinkai, Tokyo.
Nōsei Chōsa Iinkai (2008) *Kome Shoku no Sesshu Jōkyō oyobi Kōka-tō ni Kansuru Chōsa Itaku Jigyō Hōkokusho* [*Report of Commissioned Research on Status of Rice Meal Consumption and Its Effects*]. Nōsei Chōsa Iinkai, Tokyo.
Nōsei Chōsa Iinkai & Aomori Branch Office (1996) *Aomori-mai no hyōka: Bunseki to Kongo no Taiō Hōkō ni Kansuru Hōkokusho* [*Evaluation of Aomori rice: Report on Analysis and Direction of Future Responses*]. Nōsei Chōsa Iinkai Aomori Branch Office, Aomori.
OECD (ed.) (2004) *Nōgyō no Tamenteki Kinō* [*Multifaceted Functions of Agriculture*]. Translated by Sōbayashi Mikitarō. Ienohikari Kyōkai, Tokyo.
Ogura, T. (1963) *Kindai ni Okeru Nihon Nōgyō no Hatten* [*Agricultural Development in Modern Japan*]. Nōsan Chōsa Iinkai, Tokyo.
Ruellan, F. (1938) *La production du riz au Japon* [*Rice Production in Japan*]. Larose, Paris.
Saeki, N. (1986) *Kome Ryūtsū Sisutemu* [*Rice Distribution System*]. University of Tokyo Press, Tokyo.
Saeki, N. (1987) *Shokkan Seido* [*Food Control System*]. University of Tokyo Press, Tokyo.
Saeki, N. (2009) *Kome Seisaku no Shuen* [*End of Rice Policy*]. Nōrin Tōkei Shuppan, Tokyo.
Sakai, Y. (1997) *Koshihikari Monogatari* [*Tale of Koshihikari*]. Chūō Kōronsha, Tokyo.
Sakamoto, S. (1979) *Aogari Nikki* [*Diary of Prematurely Harvested Rice*]. Akita Shōbō, Tokyo.
Sakurai, M. (1989a) *Kome-Sono Seisaku to Undō Jō* [*Rice: Its Policy and Related Movements, Part 1*]. Nōsan Gyoson Bunka Kyokai, Tokyo.
Sakurai, M. (1989b) *Kome-Sono Seisaku to Undō Chu (Shōwa 22–43)* [*Rice: Its Policy and Related Movements, Part 2 (Shōwa 22–43)*]. Nōsan Gyoson Bunka Kyōkai, Tokyo.
Sakurai, M. (1989c) *Kome-Sono Seisaku to Undō Ge-1 (Shōwa 43–55)* [*Rice: Its Policy and Related Movements, Part 3a (Shōwa 43–55)*]. Nōsan Gyoson Bunka Kyōkai, Tokyo.
Sakurai, M. (1990) *Kome-Sono Seisaku to Undō Ge-2 (Shōwa 56–Heisei 1)* [*Rice: Its Policy and Related Movements, Part 3b (Shōwa 56–Heisei 1)*]. Nōsan Gyoson Bunka Kyōkai, Tokyo.
Sakuyama, T. (2006) *Nōgyō no Tamenteki Kinō wo Meguru Kokusai Kōshō* [*International Negotiation on Agriculture's Multifaceted Functions*]. Tsukuba Shōbō, Tokyo.
Sasaki, K. (1969) Buta to Sono Hinshu [Pigs and pig breeds]. *Nihon Yōton Kenkyūkai Shi* 6(3), 113–120.
Sato, Y. (2020) *Kome no Nihonshi* [*Japanese History of Rice*]. Chūō Kōron Shinsha, Tokyo.
Sekiya, S. (1981) *Nihon no Nōchi Seido* [*Farmland System in Japan*]. Nōgyō Shinkō Chiiki Chōsakai, Tokyo.

Shimizu, H. (1988) *Nihon Nōgyō no Dokujisei toha Nanika* [*What Makes Japan's Agriculture Unique*]. Nihon Keizai Hyōronsha, Tokyo.

Shimoyama, K. and Yasuda, S. (editorial supervisors) (1978) *Nihon no Nōgyō Kyōsai Hoken* [*Agricultural Insurance Scheme in Japan*]. Zenkoku Nōgyō Kyōsai Kyokai, Tokyo.

Shokuryōchō (ed.) (1992) *Shokuryō Kanri Hō 50-Shūnen Kinen-shi* [*50th Anniversary Compilation of the Food Control Law*]. Food Agency, Tokyo.

Soda, O. (1989) *Kome wo Kangaeru* [*Thinking About Rice*]. Iwanami Shoten, Tokyo.

Soda, O. (1997) *Toshi to Nōson no Ketsugō* [*Linkage Between Cities and Agrarian Villages*]. Taimeidō, Tokyo.

Suzuki, M. (2009) *Higashi Ajia Suiden Nōgyō no Tenkai Ronri* [*Logic of Development of Rice Paddy Agriculture in East Asia*]. Unpublished research report subsidized by Scientific Research Fund, Tokyo.

Taie, K. (2016) 30-nen Sanmai Karano Seisan Chōsei no Minaoshi to Kome Shijō ni Tsuite [On reevaluation of production adjustment from Heisei year 30 rice and rice market]. *Nōgyō Kenkyū* 29, 181–227.

Taie, K. (2021) Senzen no Kome no Jiyū Torihiki Jidai ni Okeru Ryūtsū Katei to Kawashimo no Shihairyoku [Distribution process during the rice free trade period in the pre–Pacific War days and influence of downstream actors]. *Nōgyō Kenkyū* 34, 177–213.

Takada, Y., Nakai, M. and Kohara, H. (2009) 1992-nen no Nōkōchi Bunpu ni Motozuku Dejitaru Nōkōchi Dojōzu no Sakusei [Construction of digital farmland soil map based on 1992 farmland distribution]. *Nihon Dojō Hiryōgaku Zasshi* 80(5), 502–505.

Takei, K. (2015) *Edo Nihon no Tenkanten* [*Turning Point in Edo Japan*]. NHK Shuppan, Tokyo.

Tanaka, J. (2021) *Yamagata-ken Sengo Inasaku-shi* [*Postwar History of Rice Production in Yamagata Prefecture*]. Ōkaze Insatsu, Yamagata.

Terayama, Y. (1979) *Ano Toki Kono Hito* [*That Time, This Man*]. Rakuyū Shobō, Tokyo.

Tōhata, S. (1966) *Atarashii Nōgyō Chitai no Dessan* [*A Sketch of New Agrarian Regions*]. Nōsei Chōsa Iinkai, Tokyo.

Tsujii, H. (1988) *Sekai Kome Sensō* [*World Rice War*]. Ienohikari Kyōkai, Tokyo.

Tsunoda, S. (1991) *Shin Mizuho no Kuni Kōsō* [*New Scheme for Mizuho no Kuni, Land of Vigorous Rice Plants*]. Nōsan Gyoson Bunka Kyōkai, Tokyo.

Unno, H. (2016) *Shokuryō mo Daijobu Nari* [*Food is Also no Problem*]. Nōrin Tōkei Shuppan, Tokyo.

Une, Y. (1984) *Gen'nōyaku Inasaku no Susume* [*Recommending Rice Production with Reduced Use of Agricultural Chemicals*]. Nisebyakushōsha, Fukuoka.

Watanabe, Y. (2016) Kono 50-nen wo Furikaette [50 years in retrospect]. *Nihon Nōgyō no Ugoki* 193, 74–96.

Yagi, H. (1983) *Suiden Nōgyō no Hatten Ronri* [*Logic of Development of Rice Paddy Agriculture*]. Nihon Keizai Hyōronsha, Tokyo.

Yaguchi, Y. (2002) *WTO Taiseika no Nihon Nōgyō* [*Japan's Agriculture Under the WTO Regime*]. Nihon Keizai Hyōronsha, Tokyo.

Yoshida, F. (2004) *Junkangata Shakai* [*Recyclable Society*]. Chūō Kōron Shinsha, Tokyo.

Yoshida, O. (2012) *Jimintō Nōseishi (1955–2009)* [*History of Liberal Democratic Party's Agricultural Policy: 1955–2009*]. Taisei Shuppansha, Tokyo.

Yoshida, T. (1990) *Kome no Ryūtsū* [*Distribution of Rice*]. Tokyo: Nōsan Gyoson Bunka Kyōkai, Tokyo.

Yoshida, T. (1993) Sanchi deno Kome Ryūtsu no Kōzō Hendō to Atarashii Taipu no Inasaku Seisansha no Tōjō [Structural changes in rice distribution in producing regions and emergence of new type of rice producers]. In: Nōsei Chōsa Iinkai (ed.) *Sanchi deno Kome Ryūtsu no Kōzō Hendō to Seisansha Jōtomai* [*Structural Changes in Rice Distribution in Producing Regions and Rice Consumers Buy Directly from Producers*]. Nōsei Chōsa Iinkai, Tokyo, pp. 1–59.

Yoshida, T. (2003) *Kome Seisaku no Tenkan to Nōkyō/Seisansha* [*Changes of Rice Policy and Agricultural Cooperatives/rice Producers*]. Nōsan Gyoson Bunka Kyōkai, Tokyo.

Yukitomo, W. (1994) Kome to Nihonjin [Rice and the Japanese]. *Mainichi Shimbun* October 28.

Zenkoku Beikoku Haikyū Kyōkai (ed.) (1973) *Beishoku to Nihonjin no Eiyō* [*Rice Diet and Nutritional Conditions of the Japanese People*]. Zenkoku Beikoku Haikyū Kyōkai, Tokyo.

Zenkoku Nōgyō Kairyō Fukyū Kyōkai/Fukyū Jōhō Center (ed.) (1978) *Suiden Riyō Saihen Taisaku no Tebiki* [*Guidebook of Rice Paddy Usage Reorganization Measures*]. Zenkoku Nōgyō Kairyō Fukyū Kyōkai/Fukyū Jōhō Center, Tokyo.

Index

Note: Page numbers in **bold** type refer to **figures**; page numbers in *italic* type refer to *tables*

Printed and bound by CPI Group (UK) Ltd, Croydon, CR0 4YY

13/09/2026

14961116-0006